Science and Engineering of Small Arms

Science and Engineering of Small Arms

Prasanta Kumar Das, Lalit Pratim Das,
and Dev Pratim Das

CRC Press
Taylor & Francis Group
Boca Raton London New York

CRC Press is an imprint of the
Taylor & Francis Group, an **informa** business

First edition published 2021
by CRC Press
6000 Broken Sound Parkway NW, Suite 300, Boca Raton, FL 33487-2742

and by CRC Press
2 Park Square, Milton Park, Abingdon, Oxon, OX14 4RN

ISBN: 9781032058245 (hbk)
ISBN: 9781032058269 (pbk)
ISBN: 9781003199397 (ebk)

DOI: 10.1201/9781003199397

Typeset in Times
by Deanta Global Publishing Services, Chennai, India

Dedicated

To

Smt. Latika Das

Her constant inspiration led to the creation and compilation of years of hard work in the form of a book. She has put up with the authors patiently and constantly inspired them to complete the work.

Contents

Foreword

Much has been experimented for the last two-and-a-half centuries in the search for a combat-worthy infantry weapon. Oceans of information on firearms have been created and are available. Still, races are on to produce the ever-eluding best weapons. It should be like that as it is aptly said that to win a war the small arms are the final weapons of victory over the vanquished. The small arms are machines governed by the laws of physics and constrained by the manufacturing technology and materials available for the construction. They are devices that call for precision and absolute reliability in action. Precise human engineering is required to achieve in the weapons the comfort of an extended body part.

To achieve so much in so little requires comprehensive ideas of both weapons and ammunition. It's heartening to note that after extensive mining of volumes of information the authors from their appreciation of the issues have attempted to present the subject matter in a style and manner best suited for interested readers. The book has been organized in 12 chapters in a logical order. It begins with the history of the development of firearms, which progressively unfolds into the connected issues and attempted solutions that led to the creation of firearms in their present form. Since ammunitions are required to be matched by the weapons, the next chapter introduces the reader to the theory of ammunition. The topics deal in a reasonable depth with the interior ballistics and give a fair idea of the development of the pressure and energy release mechanism from the propellant. The next chapter deals with the physics of small arms, which enables the user to assess different parameters of the weapon dynamics, like muzzle velocity, bullet spin, recoil energy, etc. There are clear and lucid expositions of the operating principles of automatic weapons in this section. The chapter contains an interesting derivation of the peak pressure in the barrel from ammunition and barrel parameters. The gun barrel design is a mind-blowing proposition as the authors have introduced basic machine design principles to articulate complex issues by introducing simple dynamic load factors as performed in gear design. The seemingly simple idea is an original one and may need testing for the improvement of the factors to enhance the accuracy of the calculation. The next logical element of the text is the trigger mechanism. Various possible configurations of the mechanism have been discussed and the reading will enable the reader to appreciate the appropriateness of different weapons in different situations.

The topic – anatomy of the small arms – stipulates knowledge of basic principles of weapon architecture and the discussion has been equipped with sectional views of the different weapons. The choice of representative weapons is remarkable. It includes modern weapons like AK-47, 9 mm Browning HP, S&W 53, Remington 870, HK MP 5, and FN P-90, illustrated with sectional details and specifications. The authors have well understood the role of the muzzle devices and sights in the design of an efficient weapon. A concise chapter on these devices has been aptly incorporated for the sake of completeness. The role of MIM and surface treatment in the production of small arms can be hardly understated. Both topics have been discussed in the appropriate places.

The chapter on handguns is written in detail, it is a matter of very common interest as a personal defense weapon. The block design principles have been included in this text for people who have a flair for details. The authors of the book found it fit to add a narrative on the quality of the design in firearms. The topic is contextual and a thought process ahead of time. Lastly, separate chapters on the common defects and catastrophic failures of weapons have been deliberated in detail. The narratives are invaluable treasures of years of experience. There are very few referral books in this field in India. This book would, therefore, be a boon to interested student communities and individuals who have an interest in learning more about small arms weapon systems. Any user sharing this knowledge is sure to gain immensely. Overall, the content of the book is worth reading for people dealing with the subject. I wish the authors a great success.

K Dwarakanath, IOFS
Ex-Chairman and DGOF
Indian Ordnance Factories

Preface

The invention of firearms has the greatest impact on the game of power struggles in human history. Its impact is so radical that it decided the order of hierarchy in the men as to be a ruler or being ruled. No single invention apart from firearms has changed so much the course of the history of human civilization. The history of India would have been different if the then-rulers of India in 1757 had the small arms technology. The proliferation of small arms has remained a constant worry of the world's peacemakers for long, and several treaties at the UN levels have been penned to contain the spread globally in the hands of unlawful groups. But the small arms are present in different proportions in both lawful and illegal groups and equally liked by them. So, the people, in general, may like or not, these are going to stay as long as human civilization exists. These are the weapons of final victory over the vanquished.

The challenge in the weapon design is the matching of every subsystem that makes it functional. The physical configuration is arrived at by imparting proportional functional attributes to its constituents, namely L-Lock, S-Stock, B-Barrel, C-Case, P-Propellant, B(bar)-Bullet. So, the effectiveness of the weapon ammunition system must be understood in the right perspective.

Authors have likened the problem of evaluating the effectiveness of the weapon ammunition combination to a problem of area workout of a convex quadrilateral. Each side represents a constituent firearm's component and labeled accordingly. The barrel–bullet pair has been placed on the diagonal. The overall effectiveness of the system has been viewed as the area of the quadrilateral as shown in Figure I.

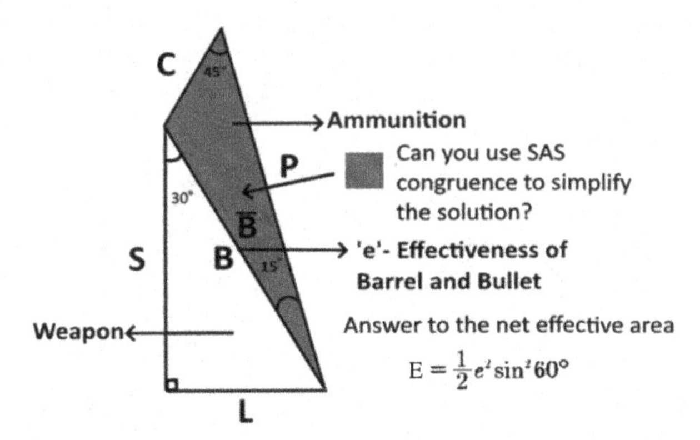

The approach to work out the net effectiveness is straight forward. Anyone with a knowledge of trigonometry and the law of cosines will naturally, add-up the area of individual triangles and arrive at the total effectiveness. But the process will be laborious and cumbersome. However, as can be seen from the diagram, by clever observation, the SAS congruence will convert the quadrilateral to an isosceles triangle and it will be easy to figure out a crisp answer of effectiveness $E = \frac{1}{2}e^2 \sin^2 60°$.

Though the technical effectiveness of the barrel-cartridge-propellant-bullet system depends on two parameters, namely piezometric efficiency which is the ratio of average pressure to the peak pressure during bullet travel, and the ballistic efficiency, which is the ratio of muzzle energy of the bullet to that of the chemical energy of the propellant, yet perfect weapon design is a situation like the quadrilateral proposition. Thus, knowing the importance, the authors have made a strenuous effort to compile relevant information and make logical deductions based on facts and their experiences for a clear understanding of the firearms. They have strategically bridged the science of ammunition with the technology of the weapons as either of these is meaningless in isolation. The book has been presented as fundamental writing for a comprehensive understanding by the layman of the overall gamut, involved in the science and engineering of firearms. It is up to the reader to pronounce the degree of success in the attempt.

Prasanta Kumar Das

Acknowledgments

The authors specifically thank Sri Surjit Das, ex Principal Director Ordnance Factory Institute of Learning, Ishapore who identified the need for such a book to train the designers in the Ordnance Factories and the maintenance crew of the users. The authors are grateful for his relentless support and encouragement. The authors acknowledge with profound regards for the time given by Shri K. Dwarakanath, the illustrious Ex-Chairman, and DGOF for his invaluable Foreword for the book. The authors also wish to thank Lt. Gen. (Retd.) Shamsher Singh, Ex-DGQA who has vast knowledge and exposure in the field of small arms for his invaluable time in summarizing the text and offering constructive ideas.

Prasanta Kumar Das

Lalit Pratim Das

Dev Pratim Das

About the Authors

Prasanta Kumar Das, M.Tech (Mech.), former General Manager RFI, OFB & IGM, Noida. He has worked in the field of small arms for more than 25 years and led important small arms projects. He has been awarded the Best Engineer's Award by the OFB and has also been awarded National Invention Award for his contribution in the small arms.

Lalit Pratim Das, M.Tech (Elec.) is experienced in the design of small arms kinematics and synthesis. He has 10 years of experience.

Dev Pratim Das, BSc-FT He is an expert on the analysis of the kinematics of small arms mechanism and design of energy flow path in the weapon system. He has also in-depth knowledge and experience of five years on the muzzle devices of various design for control of muzzleblast, flash and recoil.

About the Authors

Summary

"Small arms" are, broadly speaking, weapons designed for individual use. They include, *inter alia*, revolvers and self-loading pistols, rifles and carbines, submachine guns, assault rifles, and light machine guns. From a human security perspective, small arms are of importance. They may be used for self-defense or during the operations against the enemy. Thus, they form a very important segment of the weapon system. For the effective and smooth usage of small arms, it is important to acquire basic knowledge about their design, manufacture, maintenance, and proper preservation. Unfortunately, not much literature in consolidated form is available in India which can provide an overview of the science and engineering involved in the manufacture of small arms, how the weapons operate, and what kind of malfunctions can arise during their use. The users must collect information from various sources in different forms, collate the information to understand the basics of small arms. A need was, therefore, felt to have a book that will give an insight and deeper understanding of small arms technology.

This book is sequentially divided into various topics that deal with the different subjects of importance related to small arms. Firstly, the history of small arms is explained so that the reader gets an overview of the evolution and development process of small arms. Then the laws of physics applicable to the small arms operation are described and the different mechanics of different types of small arms are explained. Subsequently, an overview of the theory of ammunition is given. It describes different types of ammunition, design requirement of ammunition, and the types of propellant used so that the reader can have some basic knowledge as ammunition used are one of the most important parts of a small arms weapon system. The next part describes the different types of ballistics *viz.* internal, external, and terminal ballistics, that come into play when a weapon is fired.

The next section provides the details of the anatomy of a few small arms weapons which gives an idea about the structure of small arms. The advancing section gives an idea about the design concept of different parts of small arms like barrel trigger mechanism, block breech, and types of operation like gas-operated or blowback, etc. One of the sections describes the type of muzzle attachments, their requirement, functioning, and benefits. A section is also dedicated to the type of sights which can be used. In another section, the book gives an overview of the quality of design which decides the reliability and endurance of the weapon.

The book has also a part that describes the various special processes including MIM (Metal Injection Molding) associated with small arms manufacture. The surface treatment of small arms is also explained in one chapter. The common defects occurring in small arms while using them are also elaborated. Then in one part, the types of catastrophic failures that can occur due to faulty ammunition and improper use of the weapons have been presented, so that the users can take due precautions while using his weapon and avoid danger to his life and limb. A part of proof parameters is also given to give an idea about how the fitness of the weapon is tested and evaluated in a stressed condition. There is a part that describes how technical specification should be interpreted so that customers may choose the best possible weapons that can meet their technical as well as the operational requirements. Lastly, after giving all the above-mentioned overviews, the authors have also explained some of the present-day firearms in use around the world and some of the modern manufacturing technologies in the field of small arms.

At the end of the book, a set of MCQs with answer keys has been provided for self-assessment which may be of interest to the readers. Overall, the book is a valuable stimulus for the people who are interested in the field.

The author of this unprecedented knowledge bank, Shri Prasanta Kumar Das, former General Manager, Rifle Factory, Ishapore, Kolkata is a well-known figure in the field of small arms. He has undertaken this challenging task of authoring this book which is a unique piece of engineering. It will be very useful for the small arms users as well as those interested in gaining deeper insight into this subject. Shri. Prasanta Kumar Das has decades of rich experience in the field of small arms and was responsible for designing, manufacturing, and testing of the small arms of different calibers required by the defense forces, sportsmen, shooters, hunters, etc. The weapons designed and manufactured under his direct guidance are state of art and can compete with the best in the world. This book is a result of his in-depth knowledge of the subject and a keen interest in this field. I was fortunately associated with him for a brief period of two years as a quality assurance officer and had regular interactions with him. I also benefitted immensely from his knowledge and experience.

Finally, I would like to convey my sincere thanks to him for bringing out a brilliant piece of work and sincerely wish him great success in all his future endeavors.

June 2020

Shamsher Singh (Retd)
Lt. Gen. Ex-Director-General
Directorate General of Quality Assurance

Some Useful Conversion Concept and Formulae

CONVERSION TABLE

Parameters	Equivalent Values
1 mm	0.039 inch
1 Grain	0.064 gram
1 lb.	453.592 gram
7000 Grains	1 pound
The heat of combustion of typical small arms propellant	5500 J/g
1 MPa	145.038 psi
1 CUP (copper Unit of Pressure) Assault Rifle Ammunition	~1.08 psi
1 LUP (Lead Unit of Pressure) for Shotgun Ammunition	~0.95 psi
1 Joule	1 Nm
Elastic limit of typical barrel material	450–550 N/mm^2
1-foot pound	1.35582 joule
1-foot per second	0.3048 meter per second

GAUGE EQUIVALENCE

Gauge	Bore Size (Inches)
10	.775
12	.729
16	.663
20	.615
28	.550

Gauge conversion formula:

$$d_n = 2 \times \sqrt[3]{\frac{3}{4\pi}\frac{1\,{}^{lb}\!/_{n}}{0.4097\,{}^{lb}\!/_{in^3}}}$$

n = Gauge number, d_n = Bore diameter (inches)

BARREL PEAK PRESSURE CALCULATION FORMULA

$$P_{max} = C \times \frac{W}{g} \times \frac{v^2}{2AL}$$

W = Bullet Weight, v = muzzle velocity, A = Bore Cross-section Area
L = Length of Barrel, C = Coefficient of correction (3 to 4)

Coefficient of Drag C_d:

$$C_d = \frac{\text{Drag Force}}{\text{Dynamic Pressure}}$$

DYNAMIC PRESSURE

$$\text{Dynamic Pressure} = \frac{1}{2}\rho v^2$$

ρ = density of medium

BALLISTIC COEFFICIENT

$$\text{Ballistic Coefficient}_{\text{projectile}} = \frac{m}{d^2 \times i}$$

$$\text{Ballistic Coefficient}_{\text{physics}} = \frac{m}{A \times C_d}$$

where m = mass of the projectile, A = cross sectional area of the projectile, d = diameter of the projectile
i = Coefficient of form, c_d = Coefficient of drag.

CHS (Cartridge Headspace) definition:

The actual headspace of any firearm is the distance from the breech face to the point in the chamber that is intended to prevent the forward motion of a cartridge.

Definition of Ogive:

The term "**ogive**" is often used to describe the point on the **bullet** where the curve reaches full **bullet** diameter.

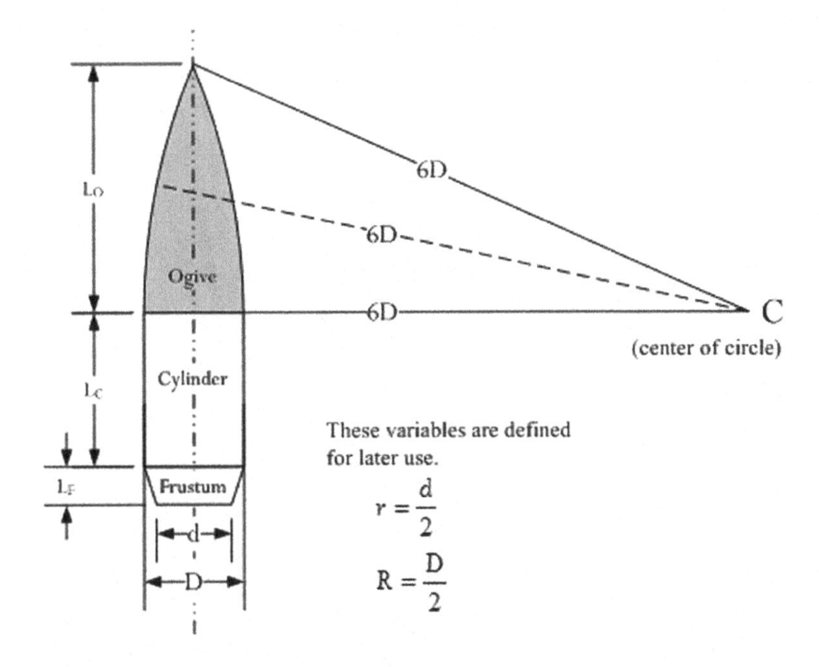

Thermodynamic Efficiency of Firearm, E

$$E - \frac{\text{Muzzle energy of bullet}}{\text{Heat of combustion of propellant}}$$

Gyroscopic Couple of Bullet

$$\tau = I \times \omega_s \times \omega_p$$

ω_s = Bullet Spin, ω_p = Bullet Precession

A few firearms terminologies:

- **Caliber**: The nominal diameter of the bore measured from land to land.
- **Centerfire**: The cartridge that is fired at the center of the base.
- **Choke**: The narrowing of the shotgun at the muzzle to reduce the spread of the pellets.
- **Magnum**: A more powerful version of the standard cartridge of the same caliber.
- **Ported barrel**: A barrel with a slot at the top near the muzzle end to reduce muzzle jump.
- **Rimfire**: The cartridge that is fired by striking at its rim.
- **Shotgun**: A smoothbore gun that shoots pellets and slugs.
- **Point blank range**: The maximum range for which sight adjustment is not required to hit a target. And the trajectory is considered almost straight.

1 A Story of the Evolution of Firearms

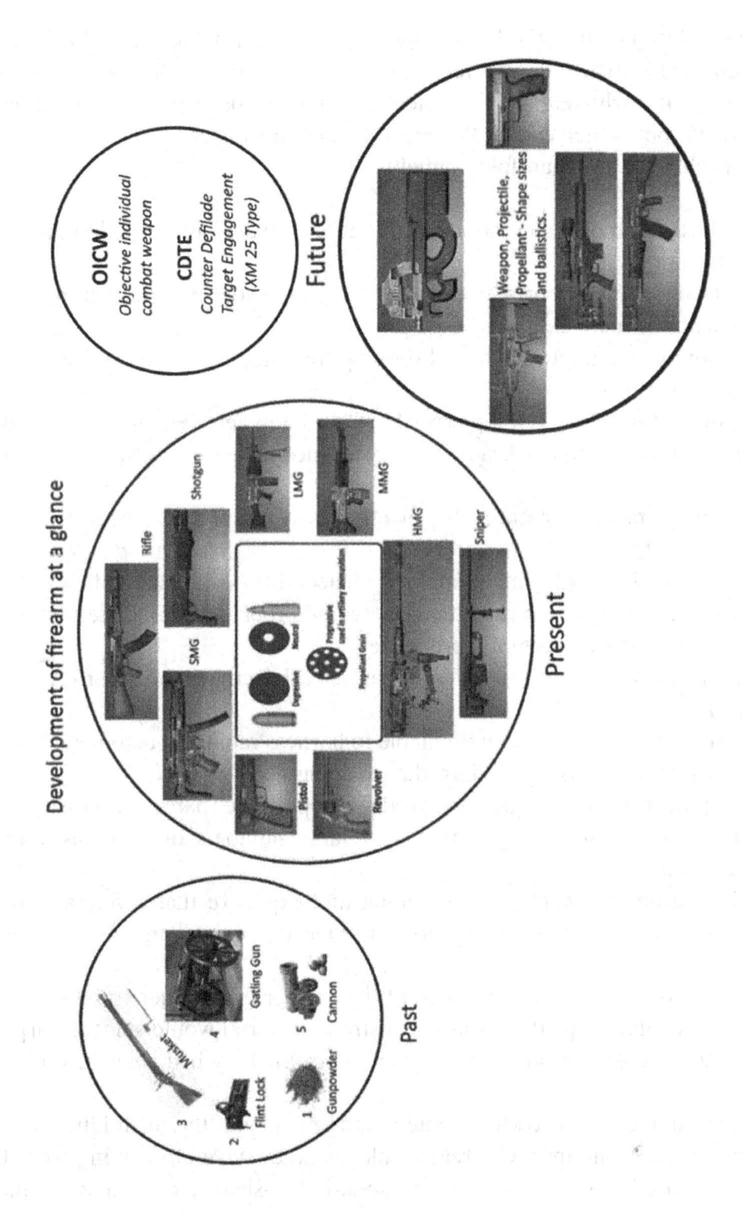

FIGURE 1.1 Development of firearms at a glance.

DOI: 10.1201/9781003199397-1

THE STORY OF FIREARMS

The evolution of firearms has been a strenuous journey since the development of gunpowder by the Chinese some thousand years ago. The Chinese could not appreciate the potential of the explosive power to exploit in the making of a weapon suitable for human needs. They mainly used the invention in pyrotechnics for recreational purposes.

During the 14th century, the knowledge of gunpowder, which was chiefly composed of sulfur, charcoal, and saltpeter mixed in some simple proportion, was passed onto the Europeans, who realized the potential of the explosive power in building up an offensive weapon which can hit the opponents from a distance.

But the problems were manifold, namely,

(1) The aggregation of the component of the gunpowder into a single maneuverable entity
(2) Identifying a suitable primer to ignite the gunpowder and integrating the same with the main explosive
(3) Designing a container which will develop the maximum explosive energy when fired
(4) Working out the design of a projectile of optimum aerodynamic shape and mass to transfer adequate kinetic energy at a desirable range with consistent accuracy
(5) Developing and understanding the mechanics to ensure the stable trajectory of the projectile
(6) Overcoming the constraints of making an accurate device that will convert the chemical energy into pressure energy and propel the projectile imparting predetermined directional properties
(7) Want of a firing mechanism that will consistently activate the firearm when needed
(8) Evolving a mechanism that will enable to harness rapid and automatic firepower as incorporated in modern-day weapons
(9) Development of techniques to produce repeatable parts en-masse, to meet the requirement of production of large numbers of weapons with interchangeability
(10) Lastly, the development of a propellant and explosive that deflagrates or detonates in the manners desired to be used as the main charge and primer

The initial solution was to find the design of the explosive container and it was soon realized that a tubular shape device made of strong material would suit the purpose. During the 19th century, progress in science and technology had been steadily taking place.

The steam engine was already invented, and as a result, the techniques of producing a cylindrical container with reasonable accuracy were already in place. The technique was used first to produce the soda-bottle-shaped cast-metal cannons which could use the spherical lead ball projectile and propel it to some 50–75 yards distance.

The canon boring techniques that were derived from the production method of steam-engine cylinders were utilized for producing the cylindrical bore with reasonable accuracy. Thus, it was the canons that emerged as the first kind of firearms to exploit the energy of gunpowder. The weapons in nature were very heavy and not portable and were primarily used for fortification against the enemy.

By this time, an empirical composition of gunpowder consisting of potassium nitrate (saltpeter), charcoal, and sulfur in the approximate proportion of 75:15:10 by weight was established as a prevailing standard. This powder was also known as "black powder", for which subsequent developments were mainly carried out by Europeans. One of the characteristics of the black powder was that only 50% of the powder burnt properly and the rest was left as a residue. This limited the muzzle velocity of the canons to approximately 1,500–2,000 ft/sec.

The drawback of this explosive was that the burning rate was invariant of the pressure and temperature. In contrast, modern-day propellants based on nitrocellulose burn at a rate almost 100–150 times faster, resulting in the higher development of pressure with efficient burnout. It was however realized by that time that neither the composition nor the grain size of the similar character of explosive is suitable for all types of weapon and the hunt for explosive of refined ballistics went on.

Till the middle of the 15th century, only the large bore cannons prevailed and the need was felt for a smaller weapon which was portable and could be employed for the offensive purpose, and thus the emergence of small arms became apparent on the horizon. The typical method of firing a cannon was very crude. First, some predetermined amount of gunpowder was rammed into the closed-end barrel over which the loosely fitted spherical projectile was placed and through a firing hole at the side of the rear, the weapon was activated. This technique was found to be suitable for use in small arms.

The initial radical changes were brought into the design of firearms through three perceptions:

(1) The propelling forces of the gunpowder could be best extracted by firing through the cylindrical hollow tube because the propellant forces on firing had to develop over space and time.
(2) The dependence of peak pressure on the inertia of the projectile and the propellant grain size had been only heuristically understood. And the rule of thumb, "The smaller the firearms, the smaller the propellant grain and projectile; and the larger the firearms, the larger the propellant grain and projectiles", has been applied in the design. This has resulted in a plethora of weapon designs of various sizes and calibers. The issue of caliber and projectile design is a matter of research and investigation even today.
(3) The fit of the projectile (spherical ball) had to be snug enough to stop the profuse leakage of propelling gas ahead of the projectile, therefore use a precisely machined cylindrical wrought-iron barrel.

(4) To give optimum ballistics (velocity and range), the projectile must be of spherical geometry. This perception, however, has been proved wrong, as can be seen using non-spherical projectile in modern weapons.
(5) The issue of managing recoil with the desired projectile energy largely remained in the empirical domain giving birth to several weapon sizes and architecture.

A scheme of the soda-bottle-shaped initial firearms (cannons) is illustrated in Figure 1.2.

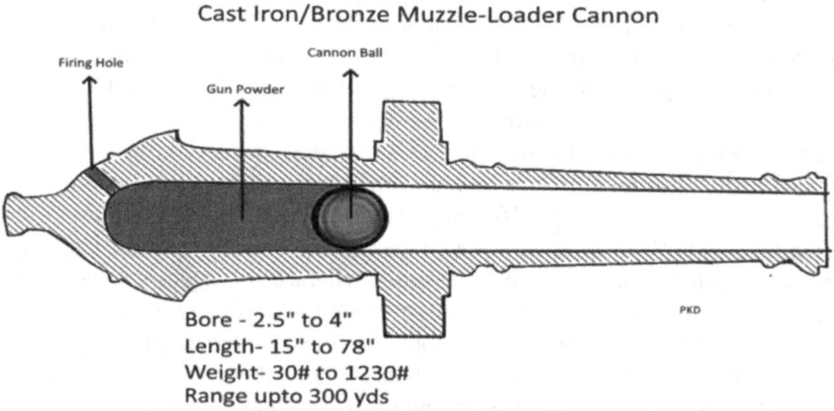

Cast Iron/Bronze Muzzle-Loader Cannon

Firing Hole
Cannon Ball
Gun Powder
PKD
Bore - 2.5" to 4"
Length- 15" to 78"
Weight- 30# to 1230#
Range upto 300 yds

FIGURE 1.2 Soda-bottle-shaped initial firearms (cannons).

THE EMERGENCE OF THE FIRST KIND OF PORTABLE ARMS

The first kind of small arms emerged in the 17th century in the form of a smooth bore muzzleloader using different types of ignition locks (firing mechanism). The development of the locks was a step forward toward the architecture of modern firearms. These muzzleloaders were also popularly known as muskets. An example of a muzzle loader is shown in Figure 1.3.

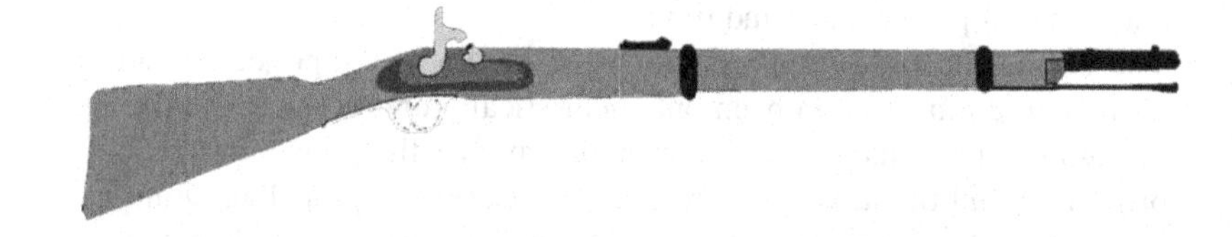

FIGURE 1.3 Muzzle loader.

The chronological order of the development of the firing mechanism used in the muskets was as below:

Matchlock
Wheelock

Flintlock
Percussion lock

Brief descriptions of the above firing mechanisms are as follows:

Matchlock – It was made in the form of an S-shaped arm, called a serpentine. It held a match, and a trigger device used to lower the serpentine, such that the lighted match would fire the priming powder in the pan attached to the side of the barrel. The flash in the pan passed through a small hole in the breech of the gun and ignited the main charge.

Wheelock – It consisted of a milled edge iron wheel, which was pressed on the fragment of flint to produce a spark on the rotation of the wheel, to ignite the powder. That was very similar to the design of a present-day cigarette lighter.

Flintlock – In the flintlock mechanism, a spring action caused the frizzen to strike the flint, generating sparks on the gunpowder in the priming pan; the ignited powder, in turn, ignites the main charge in the bore, which pushes the ball forward.

Percussion lock – Percussion lock used a self-contained, highly sensitive explosive like the modern-day mercury fulminate, or lead styphnate, which when hit sharply detonates and produces a priming flame to ignite the main charge.

In the design of muskets, the designers mainly addressed two issues, namely, the lightweight of the weapon and enhanced firing power. But because of the inevitable relationships between the recoil forces and the mass of the weapon, the designers could not do much either to enhance the firepower or to reduce the size of the weapon.

The typical weapons that were made available in the caliber of 0.69–0.75 inches were about 5.5 ft long, and weighing nearly 20 pounds. They could project fireballs weighing 2 ounces to the targets at 175 yards, that too inaccurately. Also, they required the assistance of two men to fire from a portable rest.

Some unknown genius pondered over the problem of accuracy and heuristically conjectured that spin could probably solve the problem of accuracy. So, it was seen in the 19th century that smoothbore muzzleloaders were gradually being replaced by the breechloading rifles.

INDUCTION OF RIFLED BARRELS

In lethality, smoothbore infantry muskets were relatively inefficient. These used to fire heavy spherical lead balls that could deliver bone-crushing and tissue-destroying blows as they hit the human body. But beyond 75 yards it was inaccurate. At 300 yards, balls from muzzleloaders lost most of their killing power. Well-trained soldiers could load and shoot their muskets five times per minute.

Rifled barrels, in which spiral grooves were cut into the bore, were known to improve accuracy by imparting a spin to the projectile to stabilize it.

The development of the rifled barrel, together with the introduction of percussion lock, gave birth to the architecture of the modern rifle's predecessors by the end of the 19th century. And the era of the development of modern firearms began.

MASS PRODUCTION OF THE FIREARMS

Several people contributed to the developments of firearms in the last two centuries, but it will be unfair if due credit is not given to the Europeans, who contributed most in the evolution of processes to produce firearms in large numbers and firmly established the need and designed the processes to produce standard parts and standardized patterns.

England took the first steps toward creating a national system of small arms manufacture. An ordnance office decree of 1722 laid the declaration of standard army muskets, known as "Long Land", that had a 0.75-inch caliber and a 46-inch long barrel.

The first model "Brown Bess" was the popular name given to the Long Land musket.

The U.S. government created national armories at Springfield, Massachusetts, and Harpers Ferry, Virginia, in 1794.

After the introduction of percussion ignition and rifled barrel, which occurred around 1850, the arms manufacturers all over the world started copying the American system of manufacture. This contributed to the creation of modern military small arms.

THE DEVELOPMENT OF CARTRIDGE FOR BOLT-ACTION RIFLE

With gunpowder, projectile, and the percussion locks having been in use in separate pieces, the dream of a modern form of breech loaded bolt-action rifle was very far away, till the development of improved explosive of the class of potassium chlorate and fulminate of mercury.

These were necessary for use in the percussion lock to detonate when struck with a striker. And there was also a need for subsequent integration of these percussion caps to be an integral part of a complete cartridge, which had to be an assembly of the bullet, gunpowder, and percussion caps held in a single casing of either paper or metal.

By the middle of the 19th century, the process of development of a complete cartridge was in full motion. The first such cartridges that could be successfully employed in the war was rim fired type. The fulminate of mercury was deposited in the hollow rim around the cartridge base, and by striking at the rim of the base in one spot, the round could be fired.

The Royal Small Arms Factory Enfield, in 1851 started the production of the .702-inch Pattern 1851 Minié rifle.

DEVELOPMENT OF SELF-CONTAINED BLOCK BREECH AND FIRING PIN – A LANDMARK IN BOLT-ACTION RIFLE DESIGN

To address the issue of increasing the rate of fire, it was soon realized that the problem could be resolved by loading the rifle from the breech end and firing the cartridge from behind with solid support. So, an idea was conceived that a needle housed in a steel cylinder could act as a firing pin, while the cylinder itself could support the cartridge at its base. Thus, the concept of simple bolt-action evolved.

Though the bolt action was simple in concept, it required precise workmanship. This was a remarkable feat in the rifle design.

The early cartridges were paper cartridges that posed the problem of a poor seal at the breech end. So, the advantage obtained out of the needle-shaped firing pin could not be fully exploited. The investigation of the problem gave birth to the development of center-fire cartridges of the metallic case to withstand a powerful propellant charge.

The result of the effort was the design of the 11-mm Model 1871 Gewehr (designer) in Germany and the 11-mm Modele 1866/67 Basile Gras (designer) in France. The Gewehr's rifle was a 10-shot repeater that ejected the spent case as the bolt was pulled back and fed a fresh cartridge into the chamber from a tubular magazine placed below the barrel as the bolt was pushed forward.

DEVELOPMENT OF SMOKELESS POWDER

Though the weapon came out in strong design, the gunpowder-based ammunition could not use the potential of the small arms efficiently. The black powder only produced a large quantity of solid residue, quickly fouling barrels and producing clouds of smoke, and gave only less energy to propel the projectile. In the meantime, out of the ongoing researches, nitrocellulose-based propellants were made available in the early 1880s.

These propellants burnt fully within the gun barrel, producing only gases to release the energy. So, these were named smokeless powders. These could release three times the energy of the black powder of the same mass and burned at a more controllable rate. The advantage of the design of a long cylindrical projectile over the spherical one became clear to the designers. The advent of nitrocellulose propellant and efficient cylindrical shape projectile made it possible to design small-bore firearms of about 0.30-inch or 7–8-mm caliber.

These weapons produced a muzzle velocity of about 2,500 ft/sec and could hit a target accurately at a range of 1,000 yards and beyond. It was also known by this time that beyond a speed of 700 ft/s, the lead started softening and therefore it became unsuitable for use in the high-speed projectile. So, to resolve the problem, a full-length copper-jacketed projectile was also developed, and perfected by 1881.

France was the first country to use a small-bore high-velocity rifle known as Model 1886, which was firing an 8-mm smokeless powder round. The feeding problem was resolved using a box magazine, as it was found to be very compatible with the bolt-action mechanism.

By World War II, one and all, major powers had adopted bolt-action magazine-fed repeating rifles, all of which used smokeless powder ammunition, and by this time, the development of modern bolt-action rifle was complete in the basic architecture.

AUTOMATIC WEAPONS AND THE SELF-LOADING RIFLE

The goal was achieved in designing safe and reliable firearms in the form of a bolt-action rifle, but it was found that only a very skillful firer could fire at a rate beyond 30 rounds per minute. The dreaded cavalry charge remained a reality rather than a horrible dream. To counter the threat, the hunt for a weapon that can fire and deter the cavalry attack went on.

The first outcome was the development of a semi-automatic rifle. In the U.S. John C. Garand designed the caliber .30 M1 rifle that was adopted in the army in 1936. This was a rifle in which gas was taken from a port underside the barrel and directed into a small cylinder holding a piston that was connected to the bolt through a carrier.

The gas pressure forced back the piston and the bolt, and the empty cartridge case was ejected, and the hammer was cocked. As in the present-day rifle, a spring then forced the bolt forward while the bolt picked up another fresh cartridge, and loaded the chamber in the position of ready for firing the next round.

THE SUBMACHINE GUNS AND THEIR VARIANTS

Once the sound principle of automation and the technology of ammunition were put on sound holds, the search began for the design of smaller versions to meet the emerging strategic needs. The rifles were designed for long-range, as the engagement range was around 1,000 yards during World War I. However, during World War II, the troops were required to move to accompany the armored vehicles. So they needed more portable weapons with a high rate of fire, which would be effective in close-quarter battle.

The changing scenario gave birth to the semi-automatic pistol, that used to fire a 9-mm pistol round with a muzzle velocity of 1,000 ft/sec (subsonic), and that's how the term "submachine gun" was derived.

The MP18 (Germany) was the first such successful weapon made in 1918. It had a barrel of less than 8-inch length and was chambered to fire a 9-mm parabellum round.

As the submachine gun had an effective range of only 200 yards, it became ineffective for the median military engagement range of 400 yards. So, it could not fill up the gap existing between the low-power pistol cartridge and the full-power rifle cartridge. To cover this gap, intermediate-sized automatic firearms using 7.62/5.56 ammunitions were designed and manufactured. These types of firearms popularly became known as assault rifles, and their essential features were that they had a spring-loaded magazine of 30 rounds and more, and capable of automatic fire.

Machine guns are big brothers of assault rifles. They are weapons of similar caliber as assault rifles and capable of a high rate of fire (500–1,000 rpm), with full-power ammunition. These were either magazine-fed (LMG) or belt-fed (MMG). The added problem was the management of heat generated during rapid firing. Thus, though LMGs are designed for close bolt operation, MMGs invariably have a much more complex mechanism for operation from open bolt position.

The development of machine guns was complete by the end of the 19th century. The Gatling gun and the Maxim machine guns are the earliest known machine guns. The Gatling guns and Maxim machine guns were a multi-cylinder firearm and operated by hand cranking. But now at this mature stage of development, most of the machine guns are automated either with the recoil energy or with the use of gas generated by the explosion of the cartridge.

BIBLIOGRAPHY

John Francis Guilmartin, *Galleons and Galleys*, London, UK: Cassell, 2002.

Merrill Lindsay, *One Hundred Great Guns an Illustrated History of Firearms*, New York: Walker and Co., 1967.

2 Introduction to Small Arms

2.1 PHYSICS OF FIREARMS

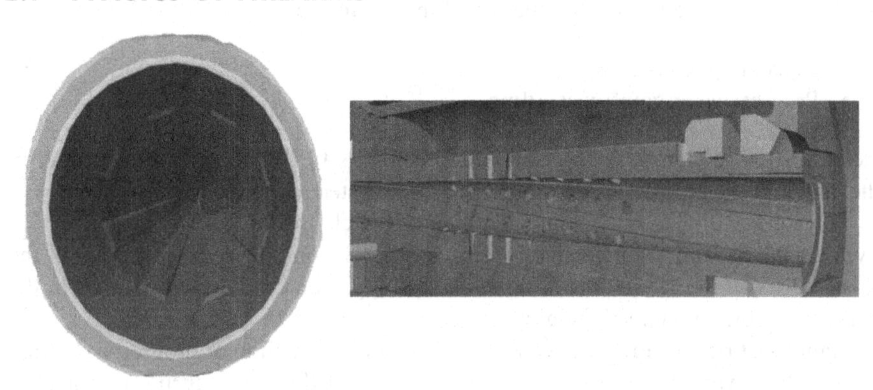

FIGURE 2.1 Rifling.

The principal function of a firearm is to throw a projectile at the desired distance with enough energy and accuracy. The obstacles in the way are the pull of gravity, the drag of winds, the force of crosswinds, and the rotation of the earth. Each of these forces attempts to deviate the trajectory from the original aims; also, in the process of the journey from gun muzzle to target, the projectile continuously loses its energy. The center of pressure of the aerodynamic forces keeps on changing, advancing toward the front of the motion, and a moment comes when the center of mass of the bullet drops behind the aerodynamic center of pressure, due to which the bullet starts tumbling. This is the general fate of all projectiles. But the shooter wants it to remain stable and accurate in the direction up to the acclaimed range of the weapon.

With a smooth-bore weapon, it is impossible to keep the projectile stable for a distance exceeding 20 yards. Rifling was the first ingenious invention to tackle the problem. The real inventor of rifling is still unknown. But some brilliant early men heuristically understood that a spin given to a projectile makes the trajectory stiffer in the travel. Thus, the idea of forcing the projectile through the spiral groove was possibly born. This spiral grooving within the barrel bore is known as rifling as shown in Figure 2.1. The number of grooves, depth, and width of the groove and the pitch of the spiral is a subject of optimization. It depends on the size of the weapon, the power of the weapon, and the properties of the bullet material.

The important parameters of the internal ballistics of the firearm, such as the peak pressure, shot start pressure, and the level of frictional forces, are dependent on the design of rifling. The rifling causes the bullet to spin at a very high speed,

DOI: 10.1201/9781003199397-2

generating the angular momentum, which essentially resists any change in direction whenever any external forces attempt to disturb the motion, by the creation of a gyroscopic couple as per the laws of physics.

The rifling is designated by twist rate, for example 1 in 12 inches, which has been explained later in the book.

Rifling Types

The design of the rifling has been investigated in detail, and several variants are in use in firearms. The following types of rifling are in use:

- Constant pitch rifling
- Progressive or gain-twist rifling

According to geometry, the constant pitch rifling is further manufactured either in the land and the groove design or as a polygon twisted along the axis.

Polygonal Rifling – Polygonal-rifled barrels are claimed to possess longer service lives for the reason, often given as a decrease in friction due to the reduction of the sharp edges of the land. This design is frequently found to be incorporated in the barrels of semi-automatic handguns.

Gain-twist rifling (Progressive Rifling) – Gain-twist rifling or progressive rifling imparts little variation to the angular momentum of the bullet during its travel of the initial few inches through the barrel, since the rate of twist is low in the region of transition from chamber to throat. This allows the bullet to stay primarily undisturbed and trued to the case mouth. When the bullet starts to bite into the rifling it is gradually subjected to more acceleration as the burning powder propels the bullet down the barrel. The gradual increase in the spin rate distributes torsion over a larger section of the barrel instead of just at the throat, where conventional rifling takes an abrupt start. This is claimed to produce lesser damage of the barrel at the throat due to frequent engagement of the bullet in the rifling during its lifetime.

Rifling Methods

Cut/Hook Cutter Rifling

The simplest method of cutting the grooves is by using a "single hook cutter". Cut/hook cutter rifling can be identified by: (i) the presence of longitudinal striations in the cut grooves and (ii) the similarity, as the same tool is used for every groove, between the micro stria in the grooves.

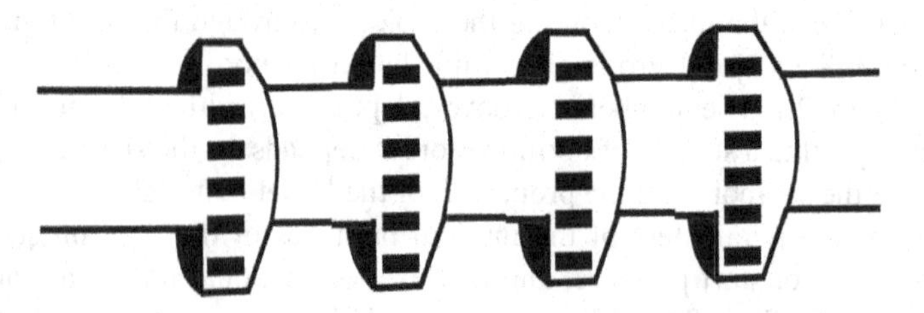

FIGURE 2.2 Broach rifling.

The broach is an integral cutter consisting of a set of grooved cutting disks (20–30 sets) which progressively increases in size. Each disk consists of the number of cutting sections corresponding to the number of the required grooves. The size of the final disk equals the size of the finished bore. When the broach is drawn through the bore with the required spiral motion, it can cut all the grooves and lands to the final dimension in a single pass. The schematic of a broaching tool is given in Figure 2.2.

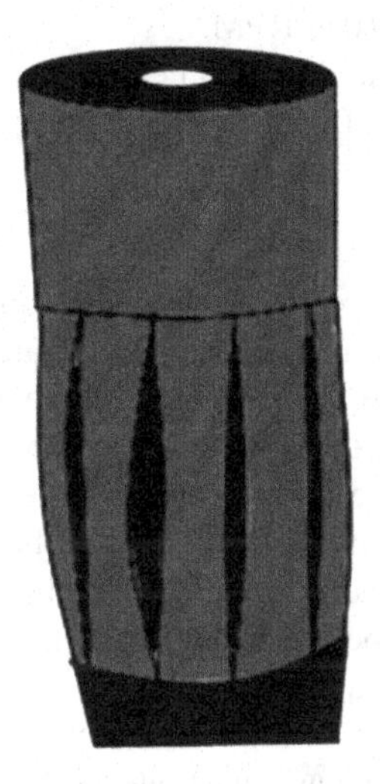

FIGURE 2.3 Button rifling.

Button Rifling

The Barrels of semi-automatics are normally within a 6-inch length. The exterior of the barrel contains several features that require complex milling. The ratio of the bore diameter to the diameter of the outside of the blank makes it unfavorable for rifling by the quick mass production process of cold swaging. See Figure 2.3.

The button rifling is a solution for mass production of handgun barrels. In this process, the barrel blank is provided with a bore that is slightly smaller than the final diameter. The button tool consists of a "Button" which is an exact negative of the rifling required. This is either pushed or pulled through the bore forcing the metal to swage out to form the spiral rifling grooves. The operation is completed in a single pass hence a cost-effective solution for handgun barrels.

Bullet Spin

A bullet fired from a rifled barrel will spin at over 150,000 RPM, which depends on the bullet's velocity and the barrel's twist rate. The general definition of the spin of an object rotating around one axis will be written as $s = \frac{v}{c}$, where v is the linear velocity of a point on the circumference of the rotating object (in units of distance/time) and c refers to the circumference of the circle on which the point lies.

A bullet that matches the rifling of the firing barrel can exit that barrel with a spin $s = \frac{v_0}{L}$ where v_0 is the bullet velocity at the muzzle, and L is the rate of twist.

For example,

- An AK47 assault rifle with a twist rate of 1 in 9.45 inches and a muzzle velocity of 2,350 feet per second will give the bullet a spin of (2,350 × 12) ÷ 9.45 = 2,984 r.p.s. or 179,047 RPM.
- The bullet material is subjected to centrifugal forces. Excessive spin can cause the disintegration of the fast-moving projectile.

Twist Rate

- The length of travel required by the bullet to complete one complete revolution around its axis is expressed as the twist rate.
- For the best performance, the barrel must have a twist rate adequate to spin and stabilize any bullet, that it'd fairly be expected to push through; however, it should not be excessive to cause a bullet fragmentation.
- Bigger diameter bullets give a lot of stability because the larger radius provides a lot of rotational inertia, whereas long bullets are tougher to stabilize, as they are inclined to be much heavier at the back and the aerodynamic pressure has higher leverage to destabilize.
- The slowest twist rates are found in muzzleloading firearms meant to shoot a spherical ball; these can have twist rates as low as 1 in 72 inches (180 cm), or slightly longer.
- The M16A2 rifle, which is meant to shoot the 5.56 × 45-mm SS109 ball, and L110 tracer bullets (NATO), incorporates a 1 in 7 inch (18 cm) or a 32-caliber twist.
- Civilian AR-15 rifles are usually found with 1 in 12 inches (30 cm) of rifling. Rifles that usually shoot longer, smaller-diameter bullets can generally have higher twist rates than handguns that shoot shorter, larger-diameter bullets.

CALCULATING THE TWIST RATE

Greenhill Formula

$$\text{Twist} = \frac{CD^2}{L} \times \sqrt{\frac{SG}{10.9}} \qquad \text{Eq 2.1}$$

where:

C = 150 (for velocity < 2,800 ft/s), and to limit the rate of spin to maintain the structural integrity of the bullet

C = 180 (muzzle velocities > 2,800 f/s)

D = bullet's diameter in inches;

L = bullet's length in inches

SG = specific gravity (Sp. Gr) of the bullet (Sp. Gr lead 10.9)

Expressing the Twist Rate in Three Different Ways

The First Method

In the first method, the twist rate is the direct statement of the result of the calculation from the Greenhill formula.

Example 2.1*

With a muzzle velocity of 2,350 ft/s, a bore diameter of 0.30 inches, and a steel bullet (Sp. Gr 7.8) length of 0.905 inches, the Greenhill formula would give a value of 12.6, which means 1 turn in 12.6 inches.

$$T_{twist} = \frac{150 \times (0.30)^2}{0.905} \times \sqrt{\frac{7.8}{10.9}} = 12.6 \quad \text{Eq 2.2}$$

This means the twist rate is 1 in 12.6 inches.

The Second Method

This describes the distance traveled in calibers or bore diameters by the bullet to complete one projectile revolution.

$$\text{Twist} = \frac{L}{D_{bore}} \quad \text{Eq 2.3}$$

where:

Twist = twist rate expressed as a multiple of the diameter of the bore;
L = the length of the twist, i.e., required to complete one projectile revolution (in mm or inches)

*Twist rate using second method in Example 2.1

$$\text{Twist} = \frac{12.6}{0.30} = 42 \quad \text{Eq 2.4}$$

D_{bore} = bore diameter (Land to land diameter, in mm or inches)

A twist rate is a simple number, and the length and bore should be dimensionally consistent.

With this method, it is not easy to understand whether a twist rate is relatively fast or slow when different bore diameters are compared.

The Third Method

In the third method, the twist is expressed by the angle of the helix made with the axis of the barrel bore and is given by

$$\text{Twist}(\alpha) = 90^\circ - \tan^{-1}\left(\frac{L}{\pi D_{bore}}\right) \quad \text{Eq 2.5}$$

*Using the third method in Example 2.1

$$\text{Twist}\ (\alpha) = 90^{\circ} - \tan^{-1}\left(\frac{12.6}{0.30\pi}\right) = 4.24^{\circ} \qquad \text{Eq 2.6}$$

The twist rate of the rifle must meet the conflicting roles of maximizing the rate of spin, consistent with bullet material, and minimizing the force of engraving while rifling the bullet. The wear and stress on the barrel are dependent on the twist rate. Therefore, the life of the barrel has a significant bearing on the choice of twist rate. In the internal ballistic, the twist rate also influences the shot start pressure. The greater the twist rate the lesser the SSP (shot start pressure).

Firearms Energy Efficiency

From a thermodynamic viewpoint, a firearm is a special type of internal combustion piston engine, where the bullet performs the function of a piston. The conversion efficiency of the energy of a firearm depends on its construction, particularly on its caliber and barrel length.

Rifle as a Machine

Can be viewed as:

a) A projectile/explosive delivery system.
b) An IC Engine.

The thermodynamic efficiency of a firearm may be as low as 25%. The overall distribution of the energy released out of propellant chemistry is shown in Figure 2.4. It can be easily appreciated that the management of waste heat and energy in the muzzle blast and flash pose considerable challenges to the firearm designer. The excellence in the design in this area rates a firearm to have a commendable attribute.

Fundamental Input of Firearm Design

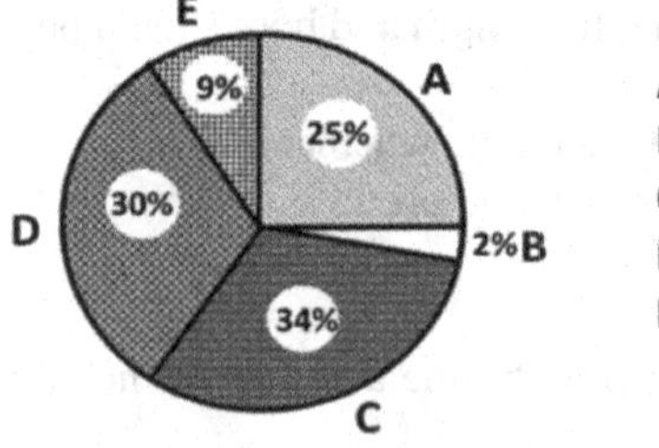

A- Projectile Motion
B- Barrel Friction
C- Hot Gases
D- Barrel Heat
E- Unburnt Propellant and shockwave

FIGURE 2.4 Energy distribution in rifle.

The pie chart as shown in Figure 2.4 provides information which is vital for the following appreciation and analysis. Given the type of propellant and the required muzzle energy, as a matter of first approximation:

1) The amount of the main charge, case volume, bullet size, barrel, and the overall size of the cartridge can be determined.
2) The heat hazard can be worked out and the necessary arrangement for heat transfer/management of the weapon system can be designed.
3) It provides the primary data for the selection and sizing of muzzle devices like flash hiders, silencers, compensators, etc.
4) The recoil can be fairly estimated and the design of the frame, recoil spring, and buffer mechanism can be worked out.
5) The distribution of energy spectra gives a primary indication of the selection of proper construction material.
6) It is the first input in the iterative process of firearm design.

Force on the Weapon System (Smooth Bore Guns)

- Assuming the gun and shooter are at rest, the force on the bullet is the same as that on the gun–shooter combination, because of Newton's third law of motion, the famous action and reaction principle. Take a system where the gun and the shooter have a combined mass M and the bullet with a mass m. On firing the gun, the two systems (M and m) separate from each other with new velocities V and v_0. Since the law of conservation of momentum states that the magnitudes of their momentum should be conserved, and the initial momentum of the system is zero, therefore:

$$MV + mv_0 = 0 \qquad \text{Eq 2.7}$$

- As force equals the rate of change in momentum and the initial momentum is zero, the force on the bullet should thus be equal to the force on the gun–shooter combination (Figure 2.5).

The Velocity of the Projectile and the Firearm

From equation 2.7, the velocity of the gun–shooter combination can be written as:

$$V = -\frac{mv_0}{M} \qquad \text{Eq 2.8}$$

This shows that though the bullet speed is very high, the recoil velocity V is very low because of the small magnitude of $\frac{m}{M}$.

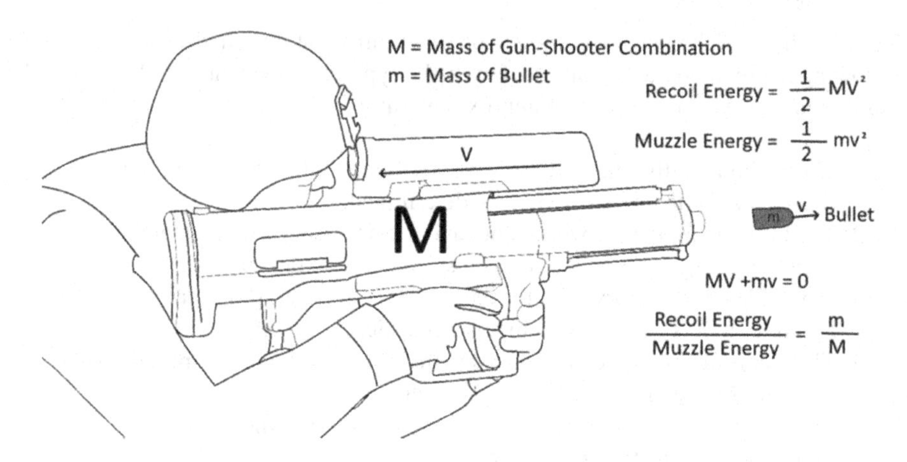

FIGURE 2.5 Gun-shooter system.

Distribution of Kinetic Energy

It is noteworthy that if the mass of the bullet is smaller than the mass of the gun–shooter system (shown in Figure 2.5), that will result in significantly low recoil energy to be transferred to the gun–shooter combination, and more will be transferred to the bullet.

The kinetic energy for the two systems is $\frac{1}{2}MV^2$ for the gun–shooter system and $\frac{1}{2}mv_0^2$ for the bullet.

The energy transferred to the gun–shooter can then be stated as:

$$\frac{1}{2}MV^2 = \frac{1}{2}M\left(\frac{mv_0}{M}\right)^2 = \frac{m}{M}\times\frac{1}{2}mv_0^2 \qquad \text{Eq 2.9}$$

If we now write for the ratio of these energies, we have:

$$\frac{\frac{1}{2}MV^2}{\frac{1}{2}mv_0^2} = \frac{m}{M} \qquad \text{Eq 2.10}$$

Energy Transfer

Because of the high-velocity and small-frontal cross-section of the bullet, it exerts large stresses in any object it hits. So, the bullet can easily penetrate any soft object.

The energy is then absorbed in the wound track formed by the bullet path. The barrels which are shorter than 10 inches may not have enough muzzle energy for causing a lethal wound channel unless fired from point-blank range with M855 ammunition.

Equation 2.10 represents the ratio of recoil energy to muzzle energy. The actual free recoil is inversely proportional to the mass (M) of the rifle for a given bullet. The shooter's upper limit of the bearable actual/free recoil energy is about 15 ft-pound and a recoil velocity of 10ft/sec. Anything beyond this limit is flinching to the shooter and uncomfortable (Figure 2.6).

Equation 2.10 is implicitly pointing out that if some portion of the shooter's mass can be reflected in the weapon system, the perceived recoil can be brought down to the shooter's comfort level. The actual recoil can be reduced through slower projectile velocity, lesser projectile weight, lesser powder charge, and higher firearm weight.

This is a simplified analysis of the recoil energy and it does not consider the recoil energy and forces generated due to the reaction produced by the high-velocity exhaust gases. The effect due to gas recoil may be as high as 90% of the recoil due to the projectile. In the recoil analysis, a suitable correction factor K_f has been considered to account for the same. The deduction is discussed below.

The net action momentum (P_a) is the summation of the momentum of the ejecta (P_e) plus the momentum of the propellant gases (P_G). If the projectile leaves with the muzzle with a velocity v_0 the propelling gas leaves the muzzle with a velocity $\alpha \times v_0$, where, $\alpha = 1.25$ for supersonic bullet (Mach > 2.5), 1.5 for Mach 2, and 1.90 for Mach ≤ 1 (for subsonic bullet).

Therefore, the net action momentum

$$P_a = P_e + P_G \qquad \text{Eq 2.11}$$

$$\Rightarrow P_a = mv_0 + m_G \alpha v_0 \qquad \text{Eq 2.12}$$

The momentum of the gun system

$$MV = (m + m_G \alpha) v_0 \qquad \text{Eq 2.13}$$

therefore,

$$V = \frac{(m + m_G \alpha) v_0}{M} \qquad \text{Eq 2.14}$$

The energy of ejecta plus gas,

$$E_{(ejecta+gas)} = \frac{1}{2}(m + m_G \alpha^2) v_0^2 \qquad \text{Eq 2.15}$$

Muzzle energy of bullet,

$$E_{muzzle} = \frac{1}{2} mv_0^2 \qquad \text{Eq 2.16}$$

Recoil energy,

$$E_{recoil} = \frac{1}{2}MV^2 = \frac{1}{2}M\left\{\frac{(m + m_G\alpha)^2}{M^2}\right\}v_C^2 \qquad \text{Eq 2.17}$$

$$\therefore E_{recoil} = \frac{1}{2}\left\{\frac{(m + m_G\alpha)^2}{M}\right\}v_C^2 \qquad \text{Eq 2.18}$$

Therefore, the ratio of actual recoil and muzzle energy from equations (2.10) and (2.18)

$$E_{ratio} = \frac{(m + m_G\alpha)^2}{mM} \qquad \text{Eq 2.19}$$

Therefore, error in the ratio from equations 2.10 and 2.19

$$Ratio_{error} = \frac{(m + m_G\alpha)^2}{mM} - \frac{m}{M} = \frac{(m_G\alpha)^2 + 2mm_G\alpha}{mM} \qquad \text{Eq 2.20}$$

Therefore, the percentage difference of estimation by equations (2.10) and (2.20)

$$d = \frac{(m_G\alpha)^2 + 2mm_G\alpha}{m^2} \qquad \text{Eq 2.21}$$

So, the correction factor K_f to be applied for recoil energy estimation from equation (2.10)

$$K_f = 1 + d \qquad \text{Eq 2.22}$$

where, m = Bullet mass, m_G = propellant mass, M = Mass of the weapon system, and α = multiplication factor. Therefore,

$$\frac{E_{recoil}}{E_{muzzle}} = K_f \times \frac{m}{M} \qquad \text{Eq 2.23}$$

Note: For rifled weapons the force, energy, and momentum equation will also contain the effect of spin of the bullet due to spiraling motion. However, these equations can be applied to assault rifles, sub-machine guns, and handguns using low or intermediate power cartridges as a measure of first approximation.

The perceived recoil is made up of the following factors

1. **The rifle fits** – The better the fit of the rifle with the shooter, the greater will be the virtual mass of the man–machine combination, hence the lesser recoil energy is perceived by the shooter. This intangible property

is responsible for the feeling of comfort in firing a bullpup design rifle, in addition to getting its desired compact feature.

2. **The muzzle blast** – The bigger the muzzle blast, the greater is the perceived recoil. The perceived recoil can be reduced by incorporating a proper compensator.
3. **Effective muzzle brake** – An effective muzzle brake will also reduce the actual/free recoil by redirecting the gases moving sideways, or rearwards but quite often increase the perceived recoil, especially when no hearing protection is worn.

The rifle fit for recoil reduction is like finding a solution of a paradoxical problem of finding A, B, C, such that the fitment of A! +B! +C! = ABC. Seemingly, the problem does not have any unique solution, but close logical reasoning yields the following unique solution: 1! + 4! + 5! = 145. Possibly similar is the thought process of matching configuration to perceived recoil reduction in the evolution of the bullpup architecture.

2.2 SMALL ARMS

Handguns (pistol/revolver)
Rifle
Machine Guns
Sniper Rifle

Typical bullet and firearm match

Weapon Type	Bullet Shape	Bullet Weight
Assault Rifle	Flat Base, SP, HP, Spitzer	53gr to 125gr
Marksman Rifle	Boat Tail, FMJ	150gr to 175 gr
Handgun	Wad Cutter, Semi Wad Cutter, Flat Nose, Round Nose	125gr to 235 gr

FIGURE 2.6 Typical bullet and firearm match.

Note: There must be a match between bullet and firearm for proper functioning of a weapon and compatibility. Figure 2.6 shows the typical bullet and firearm match.

2.2.1 Handgun

Handguns – Weapons for Self Defense

The handguns – the smallest of all firearms – are of two types:

- Revolvers
- Semi-Auto/Auto pistol

These are designed for self-defense as commonly concealed weapons because of their simple mechanics.

Handguns – Design Feature

The handgun has a short barrel and thick walls to resist high pressures. To control the muzzle jump of a handgun, extra care must be taken on account of a short barrel. The handguns too have rifling grooves like rifles to put a spin on the bullet when fired, thus increasing the accuracy and range.

Handguns are effective only for shooting at stationary targets. Each type of handgun has got a specific bore to fire only a specific caliber of ammunition. Semi-automatic pistol uses the recoil energy of one shot to reload the next round. It uses low-powered ammunition. A revolver uses rimmed cartridges and the CHS (cartridge head space) is determined by the rimmed thickness.

Auto/Semi-auto pistols use rimless center-fire cartridge, and CHS of this weapon is measured from the mouth of the case up to the back of the primer. The barrel of the revolver is no longer than 6 inches, as any increase in length beyond this doesn't enhance the muzzle velocity, because of the presence of ample clearances between the barrel and the cylinder.

Since the revolver and pistol use lead core bullets of different shapes, and either jacketed or partially jacketed, the spiral of the rifling of the barrel is of the order of 1 in 18 inches, and longer than that of the rifles. These barrels are provided with six groove rifling, and the land width is optimized to minimize the friction and force of engraving. The 0.357 MAGNUM revolver cartridges can be used in the dual role of a revolver and a rifle chambered for the same caliber.

2.2.2 Types of Rifles

Standard Rifles

- These rifles can fire a single bullet when the trigger is pulled.
- It doesn't have any semi-automatic or automatic firing modes.
- It has more power as compared to assault rifles (Figure 2.7).

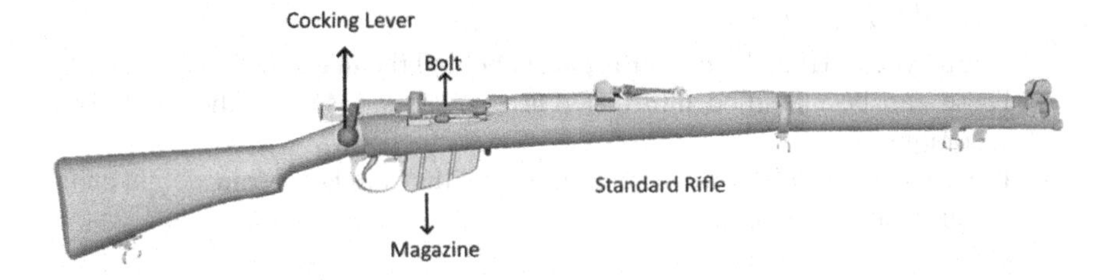

FIGURE 2.7 Standard rifle.

Assault Rifles

- These are the rifles that have different firing modes like a single, burst, and full auto shots.
- When the trigger is pulled either in burst or full auto mode it can fire multiple bullets, from a detachable 20/30 round box magazine.
- It has less power than standard rifles.
- It must have an effective range of at least 300 m.
- Multiple attachments can be done in these types of rifles.

Configuration of Assault Rifles

- Bullpup
- Conventional

Bullpup Assault Rifles

- In this type of assault rifle, the magazine is placed behind the trigger.
- This reduces the length of the weapon irrespective of the barrel length.
- These rifles are used in close-quarter combats.
- The compact and integral stock and furniture are made of thermoplastic by the process of injection molding.
- They fire intermediate power cartridge (Figure 2.8).

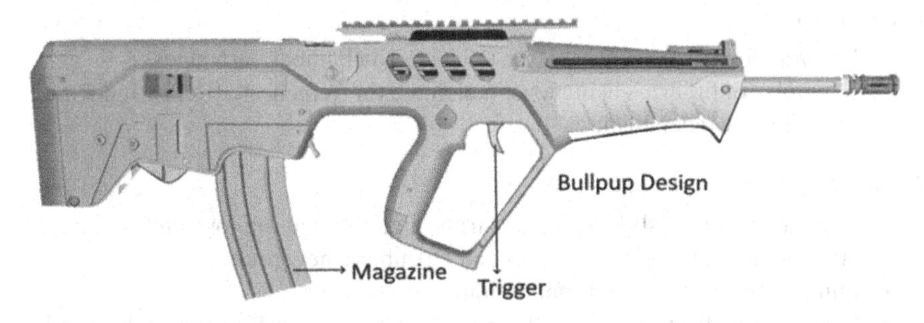

FIGURE 2.8 Bullpup rifle.

Conventional Self-loading Rifles

- In this type of rifle, the trigger is placed behind the magazine.
- These can be classified into different categories based on their type of operations.
- These rifles fire full-power ammunition and they are heavier in weight and longer in size (Figure 2.9).

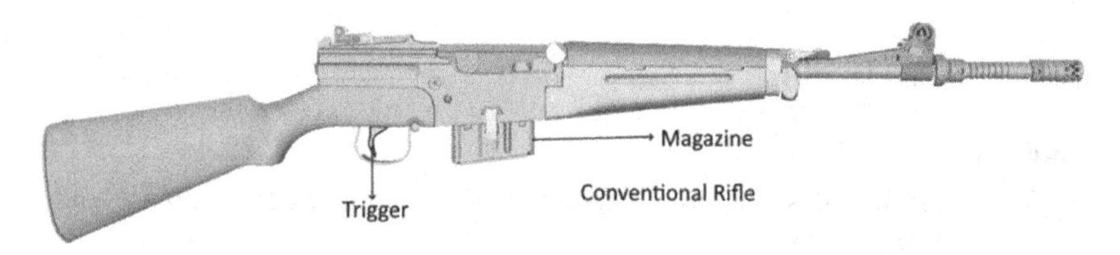

FIGURE 2.9 Conventional rifle.

Two Types of Mechanics of Operation of Firearms

A: Manual Action
B: Automatic Action

A: Manual Action

Bolt Action

- Most top-level small-bore match rifles are single-shot bolt actions.
- In bolt-action firearms, the opening and closing of the breech are operated manually by a bolt.
- Opening the breech ejects a cartridge while subsequently closing the breech chambers a new round.
- The three predominant bolt-action systems are the Mauser, Lee-Enfield, and Mosin–Nagant systems.

Lever Action

- Lever action firearms use a lever to eject and chamber cartridges.
- Examples of firearms using lever action are the Winchester Repeating Rifle and Marlin Model 1894.

Pump Action

- In pump action or slide action firearms, a grip called the fore-end is manually operated by the user to eject and chamber a new round.
- Pump actions are predominantly found in shotguns.
- Examples of firearms using the pump action are the Remington 870 and Winchester Model 1897.

B: Automatic Actions

Several operating principles are:

- Recoil operated (short recoil, long recoil), (e.g. machine guns, shotguns, and pistols)
- Blowback operated (straight, delayed)
- Blow forward operation
- Direct gas impingement
- Gas operated (long-stroke piston, short-stroke piston)

Recoil-Operated Assault Rifles

- Recoil-operated firearms use the energy of the recoil to complete the cycle of action.
- In recoil-operated firearms, only a portion of the firearm recoils, and another portion remains motionless relative to the person holding the firearm.
- The same forces that cause the ejecta of a firearm to move down the barrel also cause all or a portion of the firearm to move in the opposite direction.

Types of Recoil Operation

Short Recoil

- The barrel and bolt recoil together only a short distance before they unlock and separate. The barrel stops quickly, and the bolt continues moving backward, compressing the recoil spring and performing the automated extraction and feeding process.
- During the last portion of its forward travel, the bolt locks into the barrel and pushes the barrel back into the barrel extension (Figure 2.10).

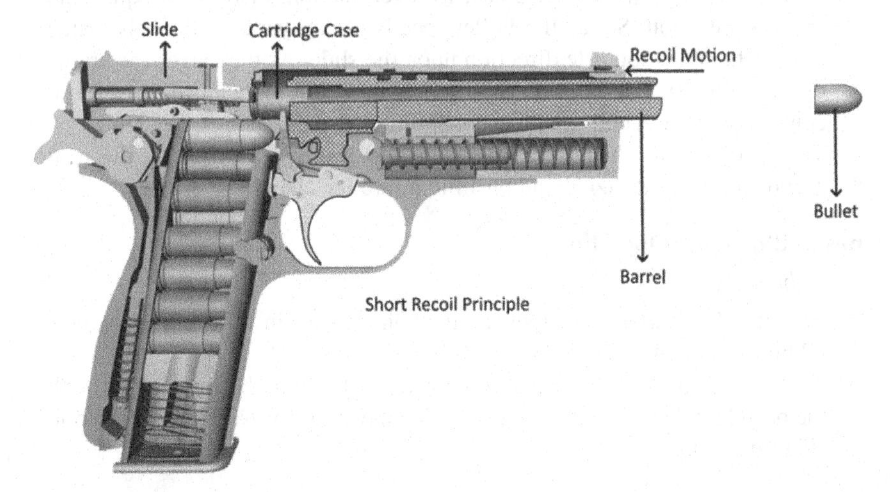

FIGURE 2.10 Short recoil.

Long Recoil

- In a long recoil action, the barrel and bolt remain locked together during recoil, compressing the recoil springs.
- Following this backward movement, the bolt locks to the rear, and the barrel is forced forward by its spring.
- The bolt is held in position until the barrel returns completely forward (Figure 2.11).

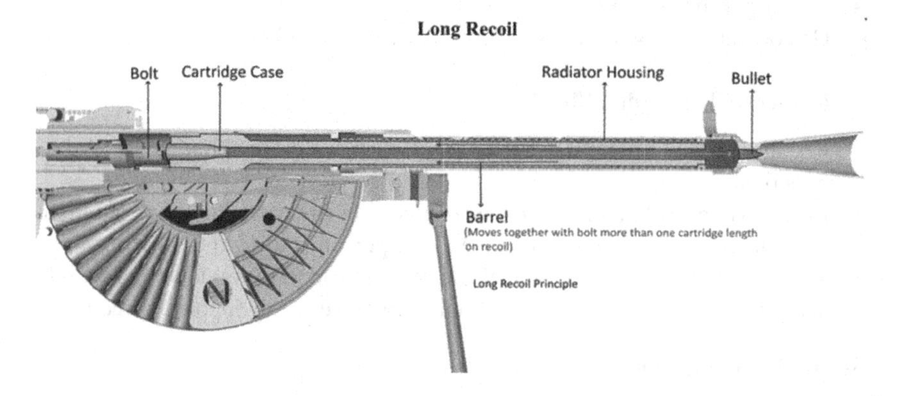

FIGURE 2.11 Long recoil-operated rifle.

Blowback Operation

- In the blowback operation, the force of the pressure is caused by detonating the gunpowder within the casing, which causes the bullet to move forward and exit the barrel.
- As a result of Newton's third law, for every action, there is an equal and opposite reaction. So, as the bullet speeds off in one direction, the same force acts in the opposite direction upon the slide. As the barrel and frame of most blowback designs are rigidly integrated, the slide moves backward, compressing the recoil spring and ejecting the spent casing.
- The recoil spring then forces the slide forward again, stripping the next round from the magazine and pushing it into the chamber.

Types of Blowback Operation

A: Straight Blowback

- In a straight blowback mechanism, the bolt rests against the rear of the barrel but is not locked in place.
- At the point of ignition, expanding gases push the bullet forward through the barrel while at the same time pushing the case rearward against the bolt (Figure 2.12).

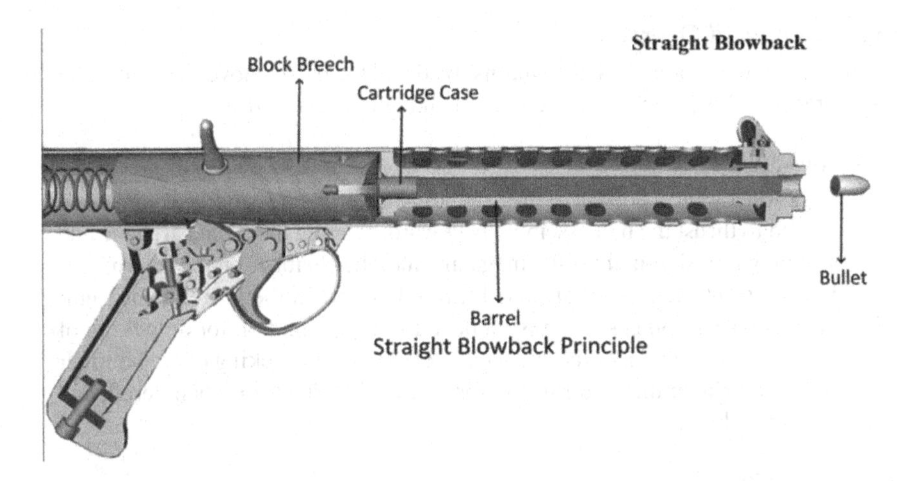

FIGURE 2.12 Straight blowback operated rifle.

B: Delayed Blowback

- In delayed blowback, the bolt is never fully locked but is initially held in place, sealing the cartridge in the chamber, by the mechanical resistance of one of the various designs of delaying mechanism.
- Types of delay mechanism – Roller delayed, Lever delayed, inertial delay, etc.
- The time of delay is a small fraction of a second. It takes a fraction of a second for the propellant gases to overcome this and start moving the cartridge case and bolt backward, and this very brief delay is enough for the bullet to leave the muzzle and the internal pressure in the barrel to decrease to a safe level (Figure 2.13).

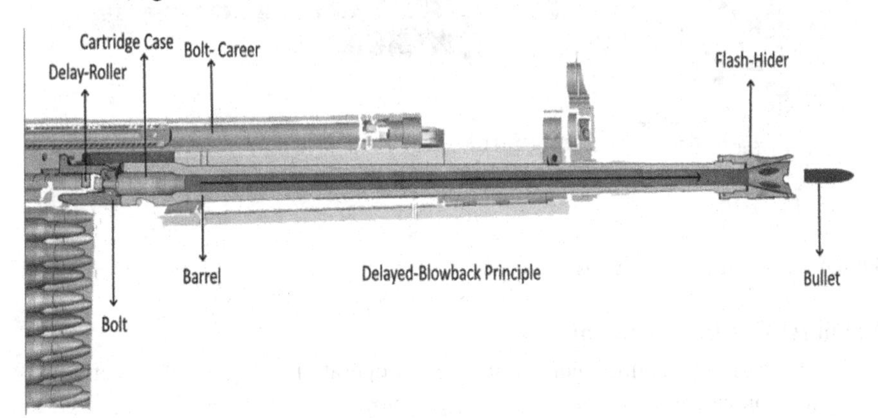

FIGURE 2.13 Delayed blowback system.

C: Blow Forward Operation

In this type of operation, the bolt remains fixed and the barrel moves forward to cycle the operation. This principle is uncommon and not generally used.

Gas Operated Assault Rifle

- In gas operation, a portion of the expanding gas from the cartridge being fired is used to thrust the bolt assembly backward, ejecting the spent cartridge and stripping a fresh round off the magazine into the chamber on the return.
- Energy from the gas is harnessed through a port in the barrel. This high-pressure gas impinges on a piston head to provide motion for unlocking of the action, extraction of the spent case, ejection, and cocking of the hammer or striker. Under the recoil spring force, a fresh cartridge is chambered, and the bolt is locked.

A: Long Stroke (LS)

- In the long-stroke mechanism, the piston is mechanically fixed to the bolt carrier assembly and moves through the entire operating cycle.
- The primary advantage of the long-stroke system is that the mass of the piston rod adds to the momentum of the bolt carrier, enabling more positive extraction, ejection chambering, and locking.
- The long-stroke gas action is smarter than the case of short-stroke gas action (Figure 2.14).

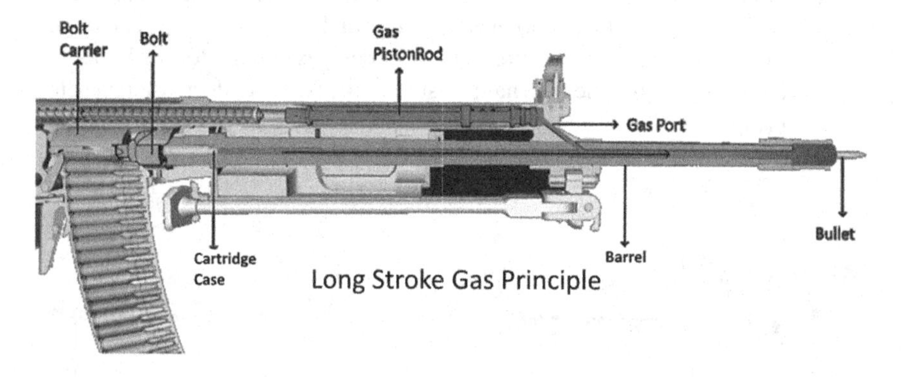

FIGURE 2.14 Long-stroke gas system.

B: Direct Gas Impingement (DI)

- In the direct gas impingement method of operation, the gas hits directly on the bolt carrier assembly through a gas tube.
- In direct impingement operation, the gas heats the receiver assembly while firing, and foul the inside of the receiver with gas residues. The bolt, extractor, ejector, pins, and springs are also heated by the high-temperature gas and reduce the operating life of these components.
- The reliability of the weapon gets reduced for continued firing (Figure 2.15).

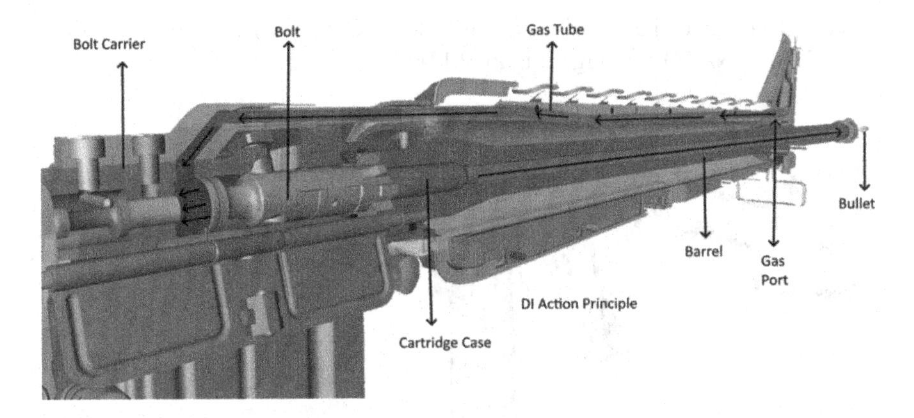

FIGURE 2.15 Direct gas impingement system.

C: Short Stroke (SS)

- In the short-stroke mechanism, the piston moves separately from the bolt group. It may directly hit the bolt carrier assembly or operate through a connecting rod attached to the bolt carrier assembly.
- In either case, the energy is imparted in a short, abrupt push, and the motion of the gas piston is then arrested, allowing the bolt carrier assembly to continue through the operating cycle through kinetic energy (Figure 2.16).

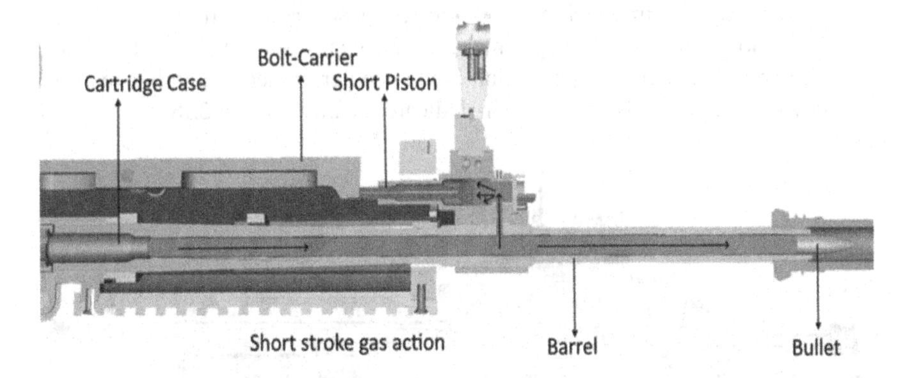

FIGURE 2.16 **Short-stroke gas system. Note:** There is the system that utilizes a muzzle cap to capture the gas after the bullet has left the barrel. This system is successful in boosting the operating power of recoil-operated guns but insufficient as a sole source of energy for generating automation as a primary operating system.

2.2.3 Machine Guns

Light Machine Guns (LMGs)

- The smallest class of machine guns is the LMG, which is designed to fire an intermediate caliber cartridge like assault rifles.

- LMGs include the M249 SAW and the MG36, both of which fire the 5.56-mm NATO cartridge (Figure 2.17).

FIGURE 2.17 FN Herstal, Belgium, M249 SAW-LMG – 5.56 NATO. Open-bolt, rear lock, gas operation, feed system – M27 linked disintegrating belt/magazine/STANAG Magazine.

Medium Machine Guns (MMGs)

- Medium machine guns are typically organized into specialized units, often consisting of a gunner and a feeder to ensure that the ammo remains properly supplied. MMG teams will usually be reasonably mobile, meaning that they can often advance with an offensive instead of after it.
- Nowadays, medium machine guns fire full-power rifle cartridges, making them more suitable for long-range use and heavy suppression. The issue with this is that their ammunition is far heavier, which is another reason why they're typically only used in dedicated teams (Figure 2.18).

FIGURE 2.18 FN, USA, M240 Bravo, 7.62mm NATO, MMG. Open-bolt, rear lock, gas operation. Feed system, belt fed.

Heavy Machine Guns (HMGs)

- The final kind of machine gun is the HMG, which often fires a .50-caliber bullet or larger. These machine guns are the least maneuverable of the bunch, and they are commonly used on vehicle mounts or by slower machine gun teams. These guns are also often found in entrenched machine gun nests.
- The advantage of a heavy machine gun is that it is equally adept at stopping enemy light vehicles as it is at halting infantry advances. This makes the HMG far more versatile in a combat role, though its heavy ammo, the weight of the gun itself, and the mount itself will often drastically reduce mobility (Figure 2.19).

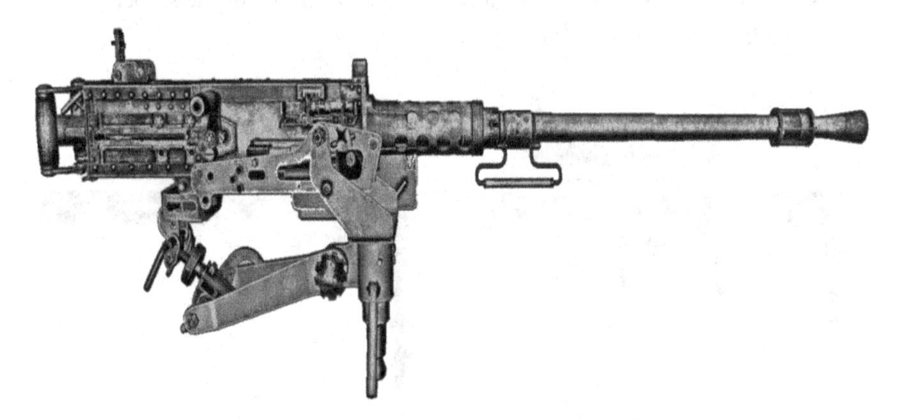

FIGURE 2.19 Browning, USA, M2. 50CAL HMG. Closed bolt, short recoil operation, belt fed.

2.2.4 Sniper Rifles

Sniper rifles are defined as long-range (600–800 m) precision rifle used to serve the role of:

1. Taking down the enemy
2. Destroying the sensitive equipment
3. Carrying out controlled detonation at long distances to destroy enemy assets

They are classed into two categories, namely Military Snipers and Police Snipers. The military snipers are of necessity of the longest possible range with an accuracy of 1–3 MOA (Minutes of Angle), while the police sniper need not have a much longer range but should be more accurate with an accuracy in the range of 0.25–1.5 (MOA).

The sniper rifle should have the flattest possible trajectory, so they are normally of either of the 0.300, 0.308, 0.338, 0.270 calibers and fires projectiles weighing in the range of 130 grs to 175 grs (grains). These projectiles have very high BCs (Ballistic co-efficient) ranging from 0.428 to 0.480. This normally fires 7.62 × 51 mm or 7.62 × 54 mm R match-grade ammunition. The legendary sniper Chris Kyle used Macmillan Tac 338 sniper for a record kill at 2,100 yds. The 0.338 snipers can hit a target casually with extreme accuracy at a range of 900 m and can harass the enemy at 1,500 m.

MacMillan Tac 50 is also a well-known sniper rifle for its three killings at the longest range.

Most of the sniper rifles are of either bolt action or semi-automatic in operation, and they are designed to fit a magazine from 5–10 rounds. A sniper normally carries 40–50 rounds on his/her duty. The small magazine of the sniper uses a short box that projects less out of the rifle, allowing a comfortable position for the sniper and thus more advantageous (Figure 2.20).

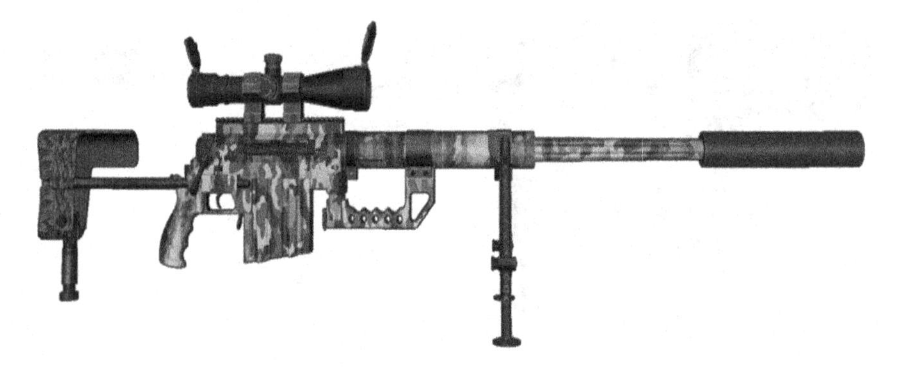

FIGURE 2.20 Chey-Tac M200 sniper rifle.

Thus, the requirement for a sniper can be summarized by the following attributes:

1. Accuracy
2. Reliability
3. Mobility
4. Concealment
5. Range assessment
6. Optics
7. Chambering for the high ballistic performance of centrefire cartridges

The noticeable characteristic features of snipers are:

1. Long Barrel
2. High magnification (4X–40X) of optical scopes
3. Specialized stocks for shooters adjustment
4. Optical/laser range finders

The distinguishing features of the optical scopes provided with the snipers are that the reticle of the scope provides the following features:

1. Drift/Deflection scale
2. Main targeting chevrons
3. Bullet drop chevrons
4. Calibration for range finding

The typical muzzle energy of different snipers may range from 3,000–30,000 J.

Distinguishing Characteristic of Sniper Barrels

- Sniper barrels are constructed of larger cross-section, and the bore manufactured to high precision to fire projectiles with consistent aerodynamic deformation and to reduce the variation in the point of impact for consecutive shots.
- The chrome plating and cold swaging introduce uncontrolled variation in the barrel bore and rifling dimensions. So, to produce an accurate barrel, rifling by precision machining is necessary and chrome plating is not recommended.
- In the assembly, barrels are kept free-floated so that the barrel contacts only at the receiver, and this minimizes the effect of the pressure on the fore-end by the slings, bi-pods, and sniper's hand, on the impact point.
- The barrel muzzle is crowned and machined to form the symmetric area at the exit to avoid asymmetry and damage and loss of accuracy.
- The spin of the rifling is important for stabilizing the projectile over its long-distance travel. Thus, keeping in consistence with the requirement of the strength of the projectile, the twist rate, i.e., normally found suitable for sniper barrels, is in the ranges of 1 in 10 to 12.
- The requirement of higher muzzle energy and complete propellant burn-out demands a barrel length of a sniper rifle of not less than 24 inches (Figures 2.21 and 2.22).

2.2.5 Shotgun

Shotgun – an important class of weapon used by both civilians and armed forces.

- Shotguns are long-barreled and smooth bored firearms. The bore is mirror-finished to reduce friction and deposits of gunshot residue to prevent corrosion. The shotguns have lower operating pressure and consequently, their barrels are of thin wall design. However, there is a limit of minimum permissible thickness mandated by the firearm regulations.
- The bore of the shotgun barrel is designed and made for a specific gauge of ammunition. The shotgun that is designed to fire a slug may have rifled barrels. Shotguns are used to typically shoot moving targets in the air in sports activities.
- In a very close-quarter battle, the semi-automatic shotguns can outperform even the best of the submachine gun.

Types of Shotgun

1. Manual
   a) Break action
   b) Pump action
2. Semi-automatic
   a) Gas operated
   b) Recoil operated

The semi-automatics are mainly used by the military and law enforcement agencies.

Calculation of the Gauge of Shotgun

The gauge number n indicates the number of spherical lead balls that constitute a one-pound mass of lead and the diameter of the individual constituent ball corresponds to the diameter of the shotgun barrel. So, a shotgun with n-gauge has a bore diameter in inches. The density of lead is taken as 11.34 g/cm³ or 0.4097 lb/in³.

$$d_n = 2\sqrt[3]{\frac{3\times 1\ \text{lb/n}}{4\pi \times 0.4097\ \text{lb/in}^3}} \qquad \text{Eq 2.24}$$

for a given d_n bore diameter in inches the gauge number *n*,

$$n = \frac{4.66}{d_n^3} \qquad \text{Eq 2.25}$$

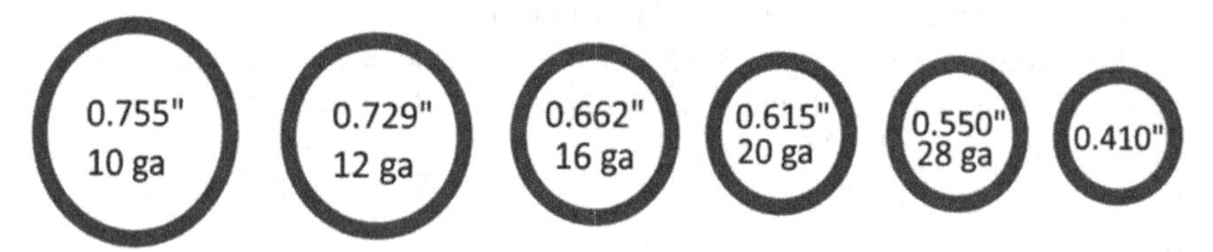

FIGURE 2.21 Shotgun bore size.

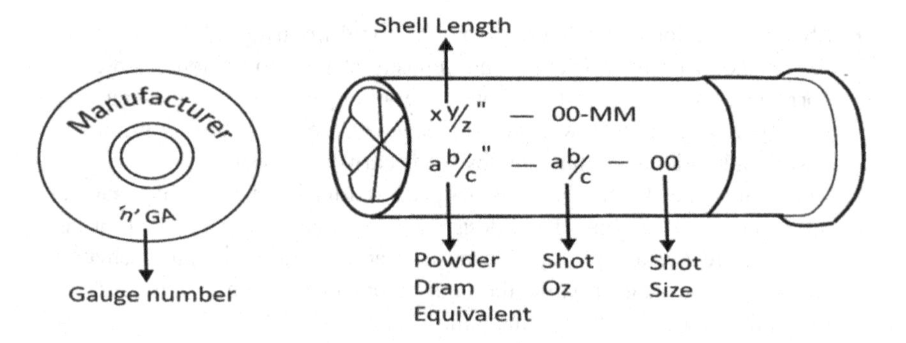

FIGURE 2.22 Shell Identification marks.

Shotgun Choke

The shotgun fires many individual shots that travel in a column. The speed of the pushing gas exceeds significantly (125–175%) immediately on the exit from the muzzle and at the same time the gas expands along the sides. Thus, it tends to spread out the shots. Therefore, a taper constriction is provided at the muzzle end of the barrel known as choke to control the spread of the shots to get better range and accuracy (Figure 2.23).

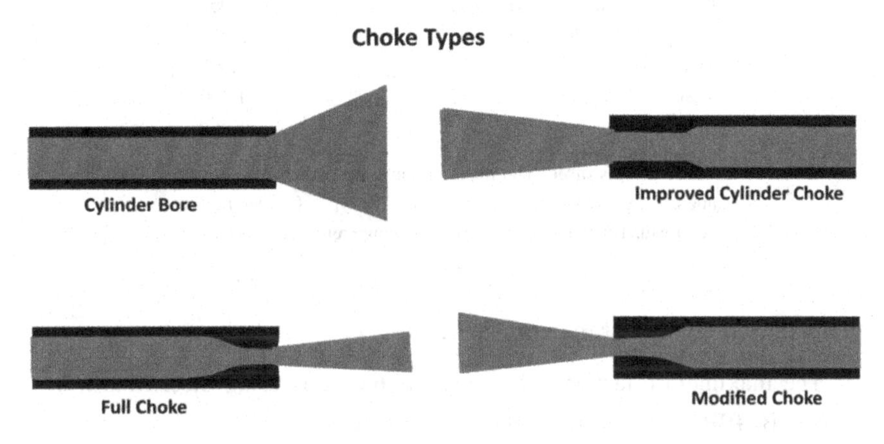

FIGURE 2.23 Choke types.

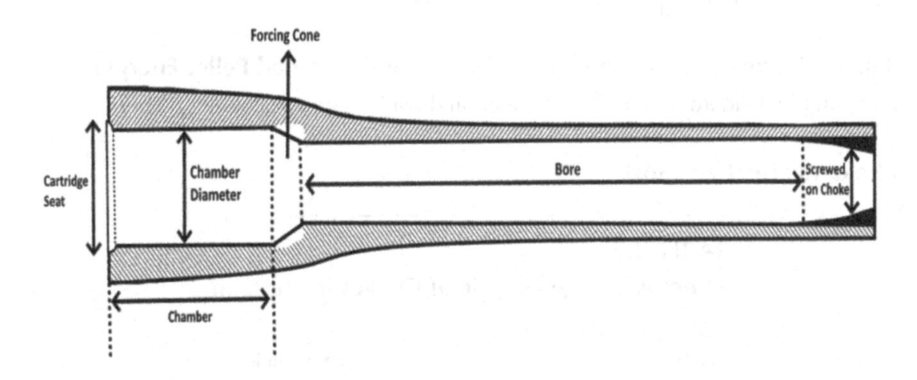

FIGURE 2.24 Basic shotgun barrel (28" Max).

Optimum Barrel Length for a Shotgun

The parameters that decide the barrel length are, reasonable weight and a well-placed center of mass that offers improved maneuverability for comfortable and efficient shooting. A shotgun with a barrel of 26–28-inch length satisfies these criteria along with the desired interior ballistics (Figure 2.24).

Approximate maximum average pressure of shotgun is given in Table 2.1.

TABLE 2.1
The Maximum Average Pressure Limit of Shotgun

Caliber	Pressure (psi)
410 Bore 2 1/2"	13,000
.410 Bore 3"	14,000
10 gauge	10,000
12 gauge (all but 3 1/2" Magnum)	12,000
12 gauge 3 1/2" Magnum	13,000
16 gauge	12,000
20 gauge	11,500
28 gauge	13,000

Note from this table: The maximum average pressure limit is the same irrespective of the case length and the type of projectile (i.e., steel, Lead, Tungsten, Slug, etc.) if the gauge remains the same.

The maximum effective range of a shotgun

- The maximum effective lethal range is 64 yards using Steel Matrix #1 #BBB, #BB* ($1\frac{1}{4}$ oz. minimum shot weight).
- The maximum effective lethal range is 70 yards using Tungsten Matrix #1 ($2, 1\frac{1}{2}$, and $1\frac{1}{4}$ oz. minimum shot weight).

Ballistic Equivalence_(Comparable Velocity and Retained Pellet Energy) for example, lead #6 shot = steel #4 shot, and lead #2 shot = steel #BB shot.

Choke vs Effective Range

TABLE 2.2
Effective Range vs Type of Chokes in Shotgun

Effective range (yards)	Type of choke
0–15	Negative or cylinder
5–20	Skeet
10–25	Improved cylinder
15–30	Light modified
20–35	Modified
25–40	Improved modified
30–45	Light full
45–60	Full

Note that for a proper shooting beyond 15 yards of range, there is a need for the progressively improved amount of choke for a change of range of every 15 yards. See Table 2.2.

Shotgun Recoil Energy

- The recoil energy of a shotgun is a mathematical function of the mass of the shot, the velocity of the shot, and the mass of the gun. Its magnitude is governed by the laws of physics. The increased amount of shot mass and shot velocity results in increased recoil. Conversely, an increase in the weight of the gun decreases the recoil energy.
- The average maximum service pressures are limited under SAAMI regulations. It may be seen that the muzzle velocity and the recoil energy are a function of shell size and shell load. In shotgun, the muzzle velocities with shell sizes of 2.75–3.0" are in the range of 1,150–1,280 ft/s depending on the load of the shell (Figure 2.25).

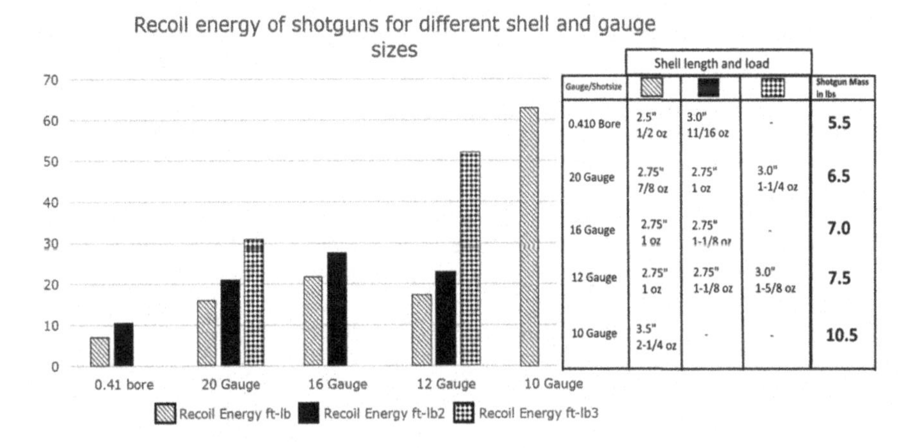

FIGURE 2.25 Recoil Energy of shotguns for different shell and gauge sizes.

BIBLIOGRAPHY

Brian J. Heard, *Handbook of Firearms and Ballistics Examining and Interpreting Forensic Evidence (Developments in Forensic Science)*, Wiley-Blackwell, West Sussex, UK, 2008.

Herman Krier, and Martin Summerfield, *Interior Ballistics of Guns (Progress in Astronautics and Aeronautics Vol 66)*, American Institute of Aeronautics and Astronautics, Inc., New York, 1979.

3 Theory of Ammunitions

DESIGN PRINCIPLES OF AMMUNITION FOCUS AROUND

- Safety
- Reliability
- Combat quality
- Weapon size and configuration

PRIORITIES IN AMMUNITION DESIGN

- The priority order of the three goals in ammunition design is as per the diagram below.
- Ammunition to resist a wide range of external environment with getting dangerous/useless to users. No failure to endanger health or life.
- Minimum failure in all-terrain and environment during its life cycle and its period of use (Figure 3.1).

FIGURE 3.1 Priorities in weapon design.

DESIGN REQUIREMENT FOR AMMUNITION

- Ensure internal ballistics for consistent pressure, velocity, spin, and muzzle blast compatible with the firearms.
- External ballistics for precise accuracy and sight calibration.
- Terminal ballistics for preconceived damage effect on targets.

DOI: 10.1201/9781003199397-3

The Requirements for Design Safety

- All materials used for production must be mutually tolerant.
- The primer should not be released during handling, loading, or shooting.
- Unintentional fall of the weapon must not cause a shot.
- Unintentional fall of the cartridge must not initiate primer and propellant charges.
- Failure probability shall be less than 10^{-6}.

CARTRIDGE

- Modern cartridges are a self-contained system (satisfying the principle of unity) consisting of a case that holds the bullet, propellant, and primer in a single unit.
- The cartridge is perhaps the single most important factor in making modern firearms practical. See Figure 3.2.
- The cartridge is a discreet mechanical unit and it seals sensitive chemicals from the external environment for a long period.

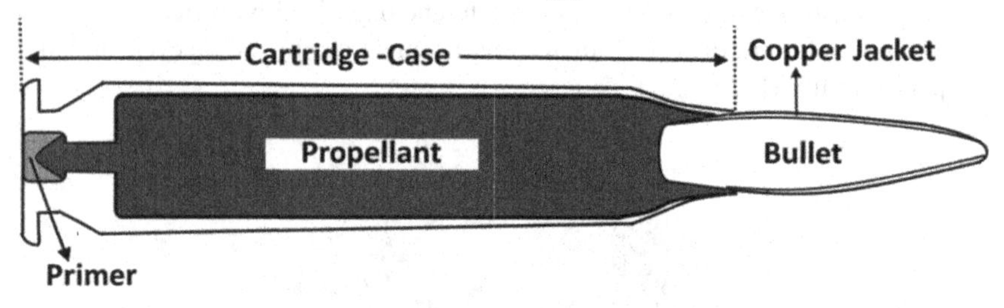

FIGURE 3.2 Cartridge components.

Components of Cartridge

- Case – shape (cylindrical/conical) and material (steel/brass) (Figures 3.3 and 3.4)
- Primer – two types of **primers** for metallic cartridges, **Boxer** and **Berdan**
- Propellant – single base/double base, grain size, mass, and volume
- Projectile – kinetic energy/incendiary/AP (Armour Piercing)/RIP (Radically Invasive Projectile), mass, and material (lead/steel/copper) (Figure 3.5).

Cartridge Specifications

Critical cartridge specifications include

- Neck size
- Bullet weight
- Caliber
- Maximum pressure
- Headspace
- Overall length

- Case body diameter and taper
- Shoulder design
- Rim type.

SAAMI (Sporting Arms and Ammunition Manufacturers' Institute) in the United States and CIP (*Commission internationale permanente pour l'épreuve des armes à feu portatives* – "Permanent International Commission for the Proof of Small Arms") in many European states decide the Cartridge specifications.

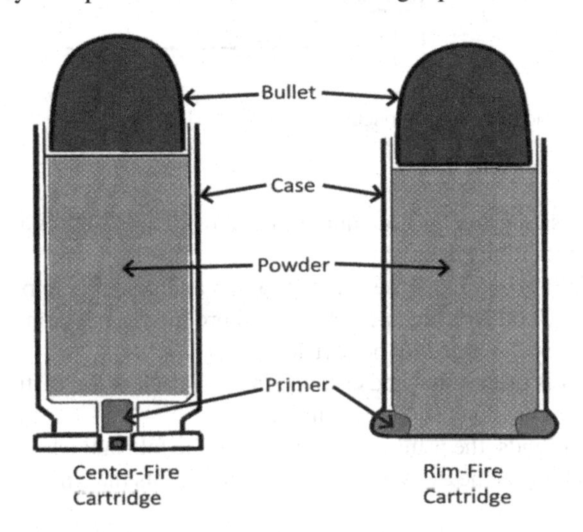

FIGURE 3.3 Types of cartridge.

Case Head Type	
Type	Image
Rebated	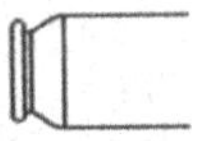
Belted	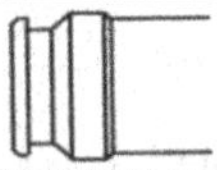
Semi Rimmed	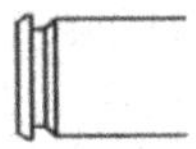
Rimless	
Rimmed	

FIGURE 3.4 Case head types.

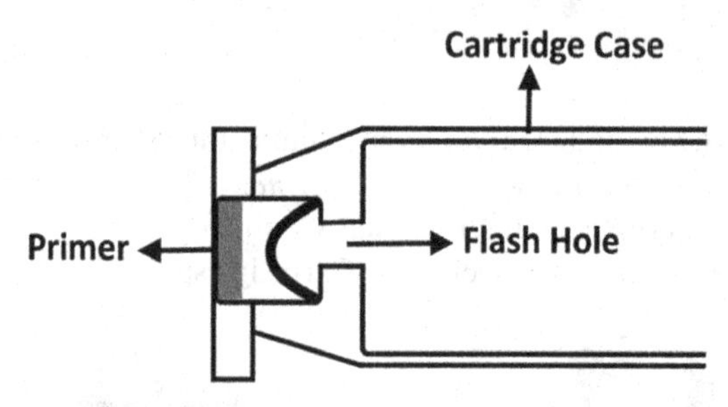

FIGURE 3.5 Cartridge case components.

Primer

- The percussion primer was an important innovation in improving the initiation of the ignition chain.
- The modern primer is in the shape of a cup and contains impact-sensitive explosives. Upon striking, it detonates and produces a hot flame to burn the main charges contained in the cartridge.
- The primer is press-fitted and crimped into the back of the cartridge case in present-day rifle and pistol ammunition.
- In rimfire rounds, the primer compound is in the fold at the back of the case which also forms the extraction rim. The striker hits the rim to detonate the primer material.

Types of Primer

Boxer Primer

A boxer primer has:

- A brass or gilding metal cup,
- A pellet containing a sensitive explosive,
- A paper disk, and
- A brass anvil.
- These parts are assembled to form a complete primer. The boxer primer is provided with a single large flash hole at the bottom of the case (Figure 3.6).

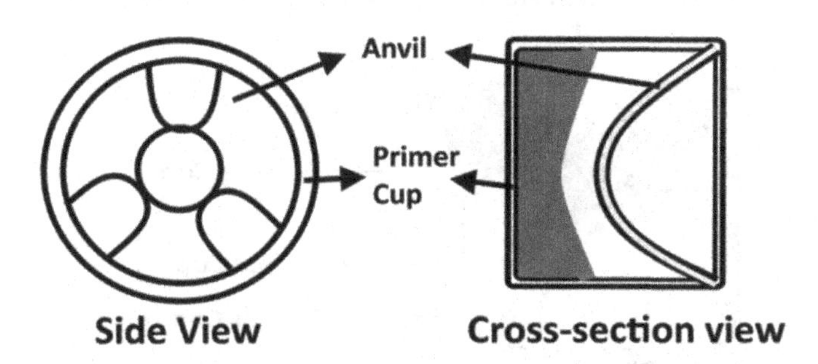

FIGURE 3.6 Boxer primer.

Berdan Primer

- In Europe, metallic cartridges are traditionally loaded with Berdan primers.
- The Berdan primer is different from the American boxer primer because it does not have an integral anvil.
- In the Berdan primer, the anvil is built inside the cartridge case and has a projection in the primer pocket.
- The primer pocket is provided with two flash holes in Berdan primers (Figure 3.7).

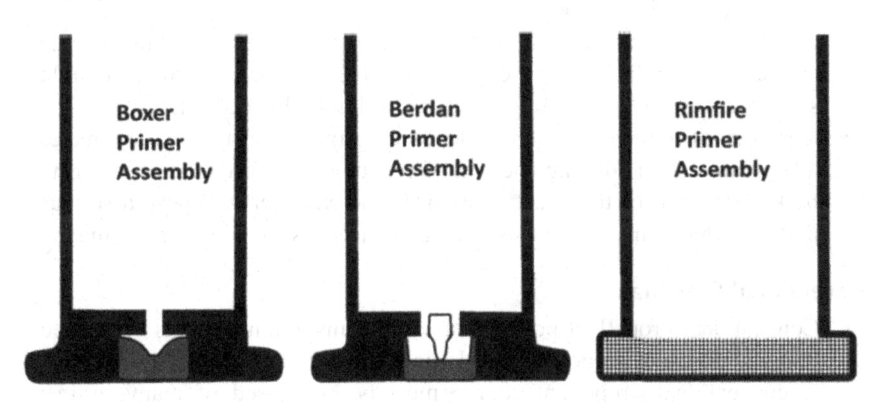

FIGURE 3.7 Different types of primer assemblies.

Battery Cup Primer

Modern shotshell primers have an additional part called the battery cup which supports the primer. The battery cup primer consists of a plain cup without any anvil which is fitted into a slightly large inverted cup (Battery cup) containing its own anvil. This self-contained assembly is fitted into a pocket in the base of the cartridge case. The battery cup may be made of brass, copper, or copper-plated steel. These primers are exclusively used for shotgun ammunition (Figure 3.8).

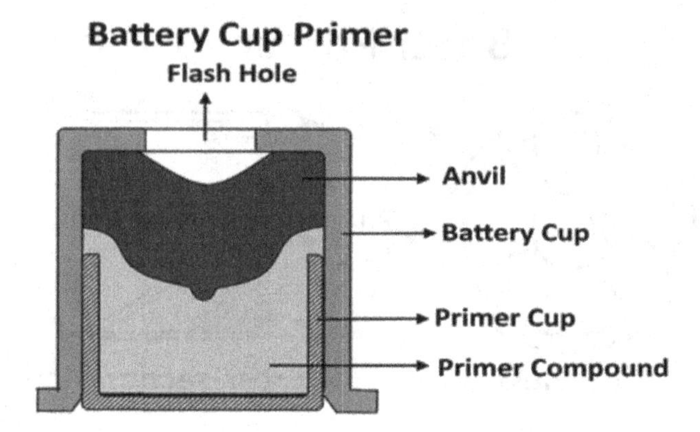

FIGURE 3.8 Battery cup primer components.

Cartridge Case Material

- Cartridge cases are, in general, made of brass or steel, but some shotgun cases are made fully of plastic with bases mostly of metals.
- The case is the carrier that holds the propellant, primer, and bullet securely. This makes it possible for the components to be transported and used as a single unit.
- The important principle is to group the relevant components as a self-contained functional unit.
- Another important function of the case is the sealing of the breach. The case expands on firing and presses against the chamber to form a gas-tight seal, stopping high-pressure gas to flow back into the receiver.
- However, since the expansion of the case against the chamber wall creates a large amount of sticking frictional force, unless the fired case can spring back, extraction of the spent cases from the chamber will pose a serious problem. Hence, the elasticity of the case material is an important parameter.

Conventional Propellant

- Conventional propellant powder for small arms ammunition is composed of many grains of propellant powder in the shape of a ball, flake, or tube.
- A conventional propellant charge must be composed of many grains; because the combustion surface thereof must be large enough to ignite it easily and smoothly and to obtain desired ballistic properties.

Propellant (Main Charge)

- The initial component of the ignition chain is the primer. Upon striking, the primer detonates (it is a primary explosive) and ignites the main charge.
- The powder then deflagrates at a controlled rate suitable to the bore diameter, projectile mass, barrel length, etc.
- The propellant is not the primary explosive, so in normal operation, it burns relatively gradually and does not detonate.

- The burning propellant generates high-pressure gases that accelerate the projectile down the barrel.
- Modern powders are produced in the shape of rods, discs, or balls and come in several sizes which together with the chemistry of the propellant affect the burning rate.
- The propellants which have smaller grains and fast burning rate are generally used in higher velocity applications and larger-grained, slower-burning propellants are used in lower velocity ammunition.
- Careful calculation and measurement of pressure versus time are needed to match the burning rates of the propellant to the sectional density of the projectile.

It is most important to keep the peak pressure within the safe pressure limits of the barrel and the rest of the system.

Modern Small Arms Propellants

- Nitrocellulose is a common origin for all small arms propellants.
- Nitrocellulose is the main energetics to release chemical energy to propel a projectile from a gun barrel.
- It is the result of treating cellulose with nitric acid in the presence of sulfuric acid.
- In a single-base propellant, the mass is softened with a mixture of alcohol and ether, which adds practically no energy to the nitrocellulose.
- In double-base propellants, nitroglycerin is added to nitrocellulose as an additional energetics. Nitroglycerin supplements the energy of nitrocellulose by the addition of chemical energy released from its component.

Smokeless Powder: The Mother of Modern Small Arms

- Nitrocellulose (detonation velocity 7,300 m/s) as the sole explosive propellant ingredient is described as single-base powder.
- Propellant mixtures containing nitrocellulose and nitroglycerin (detonation velocity 7,700 m/s) as explosive propellant ingredients are known as double-base powder.
- Alternatively, diethylene glycol dinitrate (detonation velocity 6,610 m/s) can be used as nitroglycerin replacement when reduced flame temperatures, without sacrificing chamber pressure, are of importance. The reduction of flame temperature significantly reduces barrel erosion and hence wears.
- Triple-base propellant contains nitrocellulose, nitroglycerin or diethylene glycol dinitrate, and a substantial quantity of nitroguanidine (detonation velocity 8,200 m/s). These "cold propellant" mixtures have reduced flash and flame temperature compared to single- and double-base propellants without sacrificing chamber pressure. In practice, triple-base propellants are used mainly for large-caliber ammunition.

Smokeless Propellant Components

The propellant formulations may contain various energetic and auxiliary components:

- Propellants (energetic component) – nitrocellulose, nitro-glycerin, nitro-guanidine.
- Deterrents (or moderators) – to slow the burning rate, centralities, dibutyl phthalate, dinitrotoluene
- Diphenylamine or calcium carbonate – used as a stabilizer to prevent or slow down self-decomposition
- De-coppering additives – to hinder the build-up of copper residues from the gun barrel rifling (e.g. tin, bismuth, lead metal, and compounds)
- Flash reducers – to reduce the brightness of the muzzle flash (e.g. potassium chloride, potassium nitrate)
- Wax, talc, titanium dioxide – these are additives to lower the wear of the gun barrel bore
- Other additives – ethyl acetate, a solvent for the manufacture of spherical powder; rosin, a surfactant to hold the grain shape of spherical powder; and graphite, a lubricant to cover the grains and prevent them from sticking together, and to dissipate static electricity
- The propellant shape is decided by the desired progressivity which is a measure of the burning rate. The smokeless powder has three basic compositions:
  Single base with NC (nitrocellulose)
  Double base NC/NG (nitrocellulose/nitroglycerine
  Triple base NC/NG/NGD (nitrocellulose/nitroglycerine/nitroguanidine), NGD being the third energetics
  The propellants may or may not be coated with a burning rate moderator depending on the type of application. The common propellant geometries are flake, ball, chord, single-perf, 7 perf, and 19 perf.

Degressive Propellants

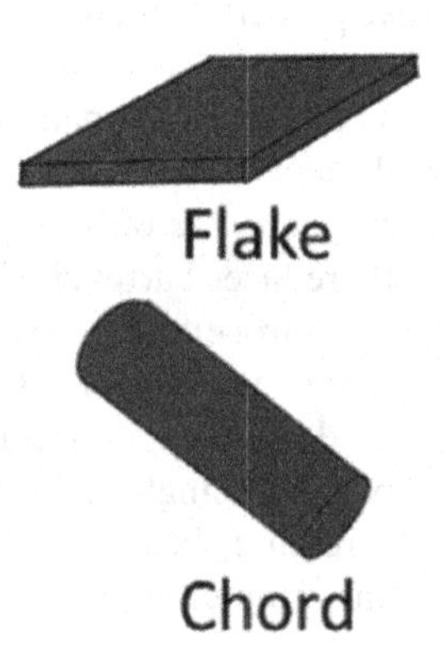

FIGURE 3.9 Degressive propellant.

Flake – flakes are selected as the propellant when rapid and complete burnout is of primary importance. The peak pressure is much lower than other high-performance ammunition as often the cases in shotgun and pistol ammunition. In these weapons, the projectile's mass/diameter ratio is low, the volume expansion will be rapid with the projectile movement, and the pressure developed falls off quickly. Therefore, in these propellants, no retardant is added to lower the pressure. These are double-base propellants.

The chord – the shape of the propellant is a right circular cylinder and is made of either single- or double-base chemistry. Their application ranges from a 5.56 × 45 mm rifle to 25 × 137 mm.

Flakes and chords are also known as degressive propellants as shown in Figure 3.9.

Neutral Propellants

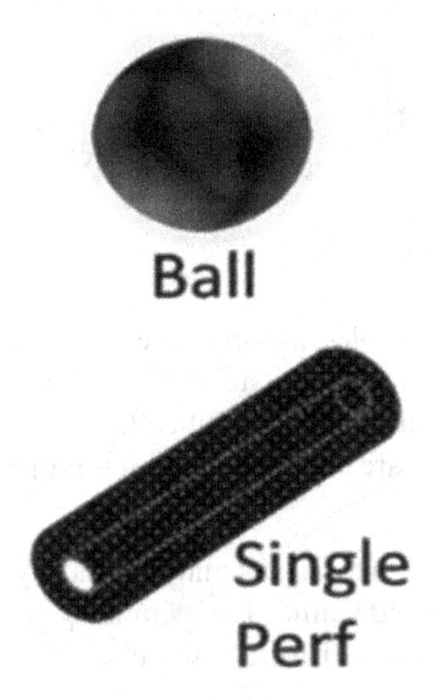

FIGURE 3.10 Progressive propellant.

Ball – "ball" is the trade name of the St. Marks Powder Company for their brand of spherical double-base propellant; however, these are not spherical but the only spheroid. Different amounts of burn rate modifiers are applied to these propellants to customize their use in ammunition that may range from 9 × 19 mm pistol to 25 × 137 mm ammunition.

Single perf (tubular) – the propellant has a hole in the center along the axis. Deterrents can be many depending on the application. It is used in all types of ammunition up to 155 mm (Figure 3.10).

Progressive Propellants

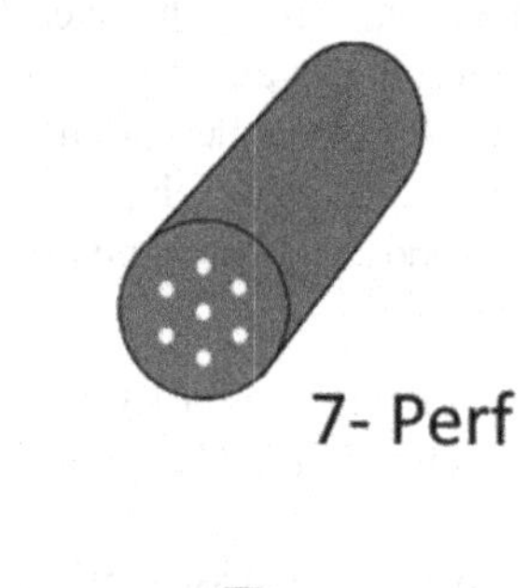

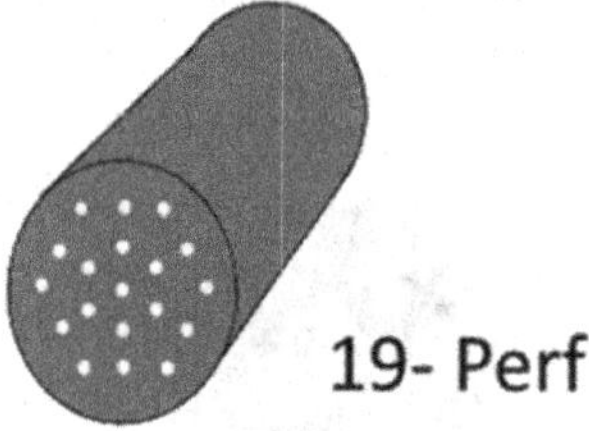

FIGURE 3.11 Progressive grains.

7-perf – These are either double-base or triple-base propellants with seven holes along the axis (one at the center and six are at the corner of the circumscribing hexagon). These are used for ammunition for caliber 25 × 137 mm up to 155 mm artillery ammunition. These powders are used for launching projectiles of the low end of the mass/diameter ratio.

19-perf – They contain 19 perforations and are mainly used for artillery ammunition in the range of 155–203 mm. The geometry is specifically designed for tank artillery caliber and is used for a low mass of the projectile by bore ratio (Figure 3.11).

Bullet Design

Bullet Shape

- The rate at which a bullet slows down in the air is determined by the ballistic coefficient (BC). The higher BC corresponds to lower retardation of the bullet and in turn, the bullet will fly a longer distance.
- The BC = SD/FF is calculated from two numbers, the sectional density (SD) and the form factor (FF). The FF is the measure of the shape of the bullet and the drag on the bullet; it varies with velocity.
- But at a basic level, it is commonsense – a bullet with a long-pointed nose, or ogive, is likely to have a better FF than one with a blunt ogive, especially at supersonic velocities.

Bullet Material and Functions

Bullets are projectiles propelled by the gun and the bullet size is expressed in terms of weight in grain and the caliber of the weapons in which the bullet is used. It addresses the following primary objectives:

Must form a shield with the gun bore without causing excessive friction.
Must fit into the rifling and retain symmetry without damaging and fouling the bore with metal deposition.
It must carry enough K.E. to the target.

The bullet shape and spin affect the aerodynamic forces on the bullet and hence the exterior ballistics. The bullet spin must be optimum to ensure proper aerodynamics and shouldn't disintegrate due to spin in the flight. Bullet shapes are optimized by considering aerodynamics, interior, and terminal ballistic requirements. Bullets are designed to perform penetrate to deform or to break upon hitting the targets thus bullet shapes are many and varied.

The bullet material strength is decided by the velocity and the spin with which, it will travel in the flight. Common bullets are made of lead, copper, or copper–nickel jacketed lead and may have a full jacket or hollow-point nose depending on the purpose. The bullets may be coated with a synthetic jacket such as nylon and Teflon. Bullets may also be made from solid mono-metal or free-cutting brass. These may be fluted in appearance for increasing terminal lethality. For avoiding fouling in polygonal rifling, hard lead alloys may be used for the bullet. The bullets may be designed and manufactured in a composite fashion for getting incendiary, exploding, armored piercing, and trajectory tracking function. Bullets are also made of powdered metals other than lead which are formed with the help of a binder and sintering. Some bullets are designed to disintegrate on the impact, these are called frangible bullets. Multiple impact bullets are on research and they have been designed as an assembly of separate slugs that are supposed to break apart in the flight in certain preconceived formations to increase the probability of heat.

However, bullet designs are controlled by several international treaties to preclude the objectionable impact on the target.

Classes of Bullets

Maximum accuracy bullet
Maximum damage bullet
Expanding or fragmenting bullet

Ballistic Coefficient Calculation

The efficiency of the bullet to retain the kinetic energy is expressed in comparison to the ballistic coefficient of the first reference (G_1) or (G_7) which are standard shape projectile.

The ballistic coefficient in any other shape is calculated as follows:

$$BC_{Projectile} = \frac{m}{d^2 i}, \quad \text{Eq 3.1}$$

where d = Bullet diameter, m = bullet mass, I = coefficient of form.

and,

$$i = \frac{2}{n}\sqrt{\frac{4n-1}{n}}. \quad \text{Eq 3.2}$$

where n = number of the caliber of the projectile ogive.

For high-drag bullets FF ≥ 1.0.

Example:

For 0.30–06, 180gr G_1 bullet ogive, R=2d, for n=2, where 'R' is the radius of the ogive and 'd' is the diameter of the bullet, therefore

$$FF = \frac{2}{2}\sqrt{\frac{(4\times 2)-1}{2}} = 1.871$$

to express BC in lb in^{-2}, $BC_{Projectile} = \frac{m}{7000\, d^2 i}$, where d = bullet diameter, m = bullet mass, I = *form factor.*

Therefore, $BC_{G_1} = \frac{180}{7000\times(0.30)^2\times 1.871} = 0.153$

Since this is a low BC, very sharp-pointed bullet form is used for 180 gr bullet for which FF is 0.6, and this value will result in a BC of 0.476 (Figure 3.12).

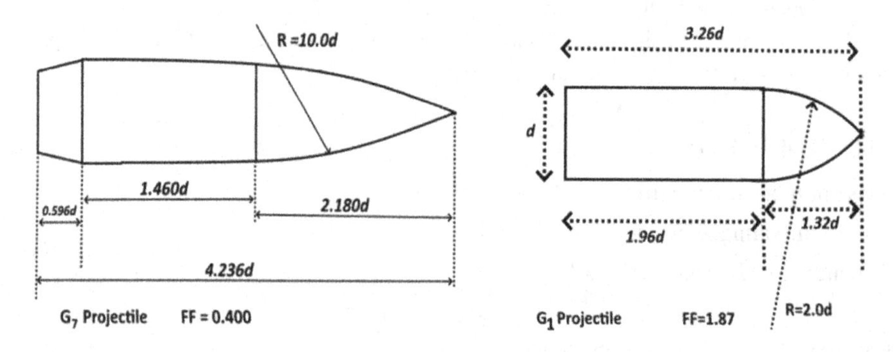

FIGURE 3.12 The ballistic coefficient is evaluated in terms of reference bullet G_1 and G_7,

Shape	*Typical form factor FF*
 Very Sharp	*0.6*
 Moderately Sharp	*0.8*
 Moderately Blunt Profile	*1.0*
 Semi-Wad cutter	*1.60*
 Hollowpoint Semi- Wad cutter	*1.80*
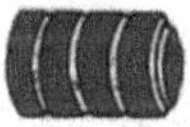 Wad cutter	*2.00*

FIGURE 3.13 Bullet shapes.

Note: The FF depends on the velocity of the bullet in addition to the geometry. Figure 3.13 is only to give a conceptual idea of the bullet shape in the calculation of the ballistic coefficients (Figure 3.14). The bullet may be full metal jacketed (FMJ) or total metal jacketed (TMJ).

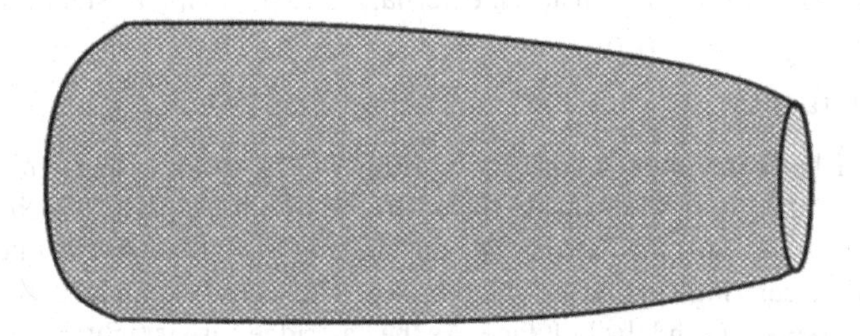

FIGURE 3.14 Frangible bullet. The bullet may be full metal jacketed (FMJ) or total metal jacketed (TMJ).

Effect of Bullet Shape

- A well-shaped bullet gives the options that a typical NATO bullet shape does not have.
- Keeping the bullet weight the same, improved long-range performance can be achieved.
- If range extension is not the primary aim, bullet weight can be reduced while keeping the same BC (Ballistic Coefficient) as the NATO bullet, thereby reducing cartridge weight.

Bullet Weight

- If the bullet weight is reduced, the next choice is the propellant load and cartridge case that can be left the same, then a higher muzzle velocity and flatter trajectory can be obtained.
- Or the MV (muzzle velocity) can be kept the same but the propellant load and size of the cartridge case can be reduced, gaining further reductions in recoil and ammunition weight.
- These are several choices to have, but these choices can be only made if a well-shaped bullet is adopted and a high ballistic coefficient is built-in by selecting bullets with long ogive.
- The peak pressure developed in the barrel is proportional to the bullet weight. An excessively high grain of bullet not only increases the recoil but can build-up excessive peak pressure causing barrel bulge, or a catastrophic failure.

3.1 INTERNAL BALLISTICS

Ballistics

Ballistics is the science of propulsion, flight, and impact of projectiles. It is divided into four sections: Internal, transition, external, and terminal ballistics respectively.

Internal Ballistics

The internal ballistics begins with the ignition of the primer. The ignition of the primer causes the gunpowder inside the cartridge to burn rapidly, which builds the internal pressure against the inertia and resistance of the bullet. As the internal pressure within the cartridge increases, the brass cartridge case expands, until it encounters the chamber walls and the bolt face. As the cartridge case is tightly seated against the chamber, the increasing gas pressure created by the burning powder forces the bullet out from the cartridge case, and through the barrel which blows out the bullet like a vented pressure vessel. And the speed with which the bullet leaves the muzzle is called the muzzle velocity.

The movement through the barrel is the study of internal ballistics. The bullet is stabilized in the flight due to its spin, as the result of its spiral movement through the rifling in the barrel. The spin creates a gyroscopic action, which gives rise to a restoring torque when any external force tends to change the direction of the motion. The rifling of the barrel creates this spin. The spin can be clockwise and counter-clockwise. While the spin stiffens the trajectory, it creates a left or a right drift of the trajectory due to the bullet's yaw of repose due to gravitational pull.

Rifling twist rate is measured by one complete rotation of the bullet in a prescribed number of inches of barrel length. For example, a rifle barrel that causes the bullet to rotate 1complete turn in 12 inches is referred to as a 12-inch twist, or 1:12. A Colt M-16/AR-15 has a 1:12 (slow), 1:9 (medium), or a 1:7 (fast) twist rate depending on what type of bullet the barrel was designed to shoot. Different bullet calibers and weights may require different twists and velocities to stabilize them; for example, a 1:12 twist barrel may not stabilize bullets heavier than 60 grain, causing these bullets to yaw or wobble making them inaccurate. The faster the twist the faster the bullet will spin; longer bullets need to be spun faster to ensure stability.

When the bullet is propelled from the cartridge case and impacts the rifling, a series of vibrations begin. The bullet initially strikes the rifling causing a Longitudinal Vibration or end-to-end vibration within the barrel. This is when the rifle recoils, moving back into the shooter's shoulder. When the rifle can't move any farther to the rear, the barrel will start to rise upward, bowing in the middle of the barrel causing an up-and-down movement of the barrel. This whipping motion is known as Vertical Vibration. There will also be a Lateral Vibration, or side-to-side movement, of the rifle barrel. As the bullet is forced down the barrel, the rifling imparts a spin to the bullet causing a Torsional Vibration to the barrel. All the four vibrations that occur at the same time are called a Barrel Whip.

To ensure absolute accuracy, the barrel must vibrate in the same fashion every time the rifle is shot. This accuracy can be accomplished by using good quality, heavy rifle barrels, good quality ammunition, and consistent shooter's skill. However, it is important to note that the barrel should not touch any other part other than the receiver. A barrel that is not touching the stock is called Free-Floating.

Factors in Internal Ballistics

- Load density and consistency
  Load density is defined as the percentage of the space in the cartridge case which is filled with propellant powder. In general, loads close to 100% density burn more consistently than lower-density loads.
- Chamber
  Straight vs. bottleneck, the large diameter in straight geometry allows a short, stable bullet with high weight, and the maximum practical bore volume to extract the most energy possible in each length barrel. But these pose feeding problems in a semi-automatic or automatic weapon.

Aspect Ratio and Consistency

When a rifle cartridge is selected for maximum accuracy, a short, fat cartridge with very little case taper may render higher efficiency and more consistent velocity compared to a long, thin cartridge with a high amount of case taper (the reason for a bottle-necked design).

Friction and Inertia

- As the rate of burning of smokeless powder varies directly with the pressure, the pressure build-up initially (i.e. "the shot-start pressure"), has a significant effect on the final velocity.
- In small caliber firearms, the friction holding the bullet in the case decides how soon after ignition the bullet starts moving, with the motion of the bullet the swept volume increases causing a drop in the barrel pressure. The friction adds to the inertial resistance of the bullet movement and hence affects the acceleration of the bullet so the difference in friction in bullet movement can cause a change in the slope of the pressure curve.
- In general, a tight fit is desired, to the extent of crimping the bullet into the case, sizing the case to allow a tight interference fit with the bullet, can give the desired result.
- The bullet must tightly fit the bore so that the high pressure of the burning gunpowder gets sealed. This tight fit produces a large frictional force.

The Role of Inertia

- The acceleration is of the order of tens of thousands of gravities, so even a light projectile as light as 45 grain (2.9 g) can provide over 1,100 newtons (247 lbs) of resistance due to inertia.
- Variations in bullet mass, therefore, have a great impact on the pressure curves of smokeless powder cartridges.

Typical Pressure–Velocity Relationships (5.56 mm 55 gr Round from 20-Inch Barrel)

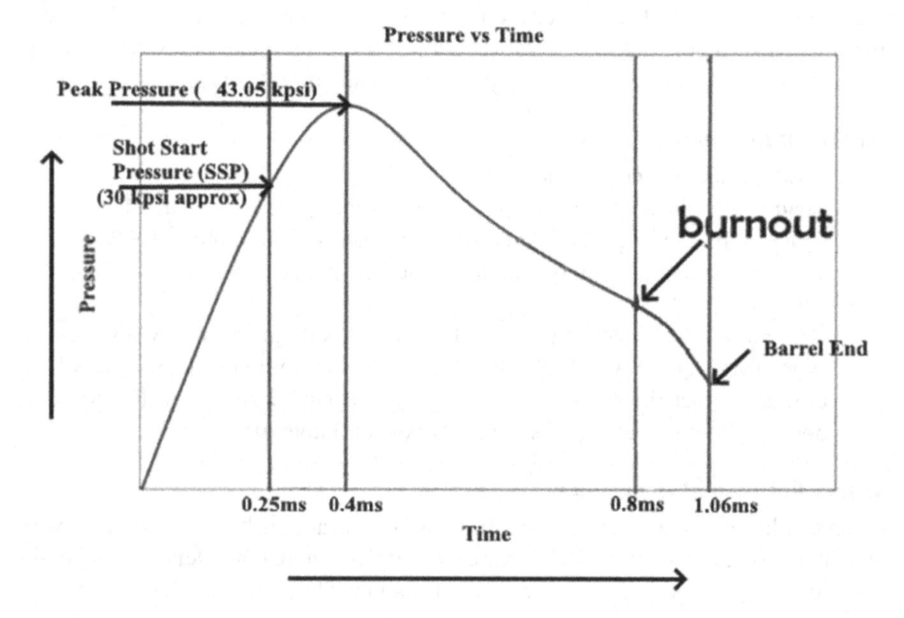

FIGURE 3.15 Pressure vs time.

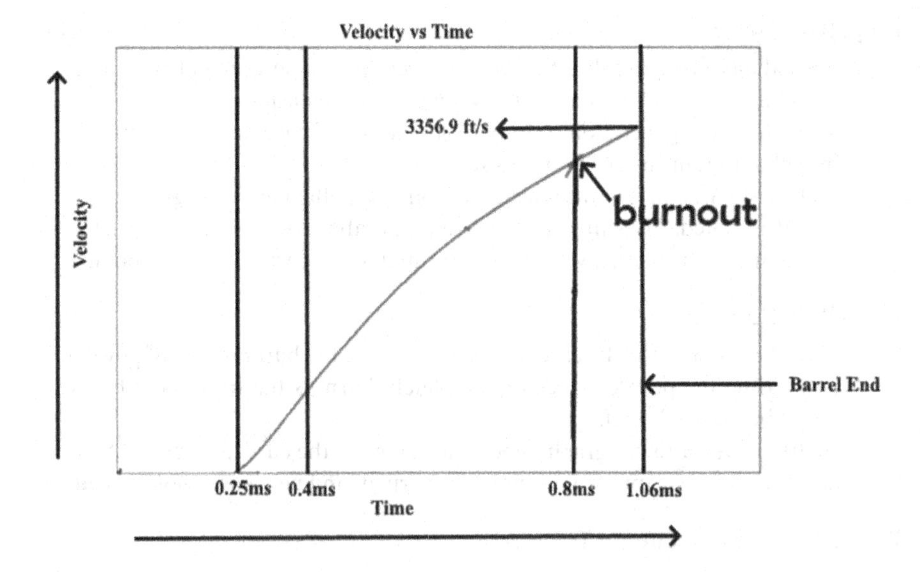

FIGURE 3.16 Velocity vs time.

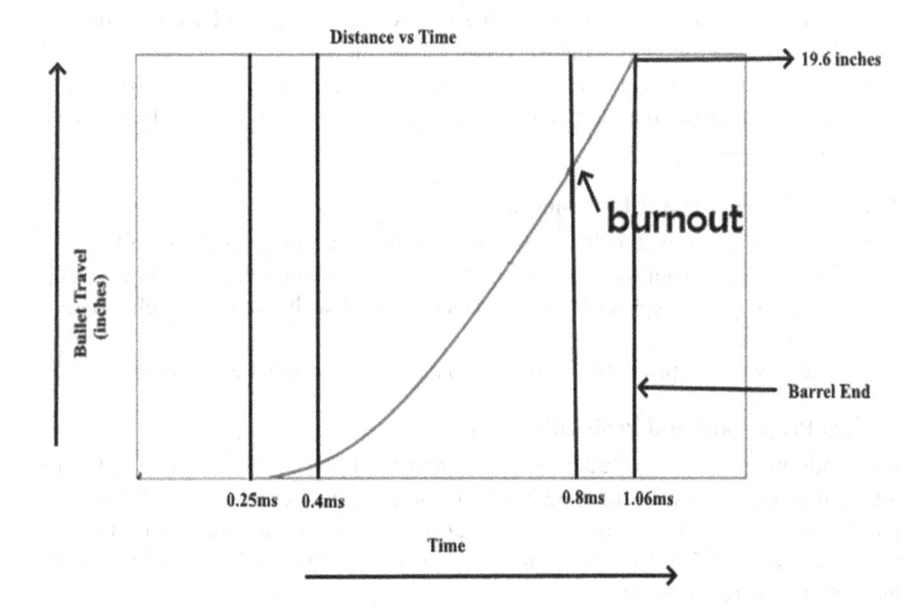

FIGURE 3.17 Distance vs time.

This is a graph of a simulation of a 5.56 mm (55 gr) round being fired from a 20-inch barrel (Figures 3.15–3.17).

Propellant Design

- Propellants are carefully matched to firearm strength, chamber volume, barrel length, and bullet material, weight, and dimensions.
- The rate of gas generation is proportional to the surface area of the burning propellant grains as per Piobert's law.
- Piobert's law applies to the reaction of solid propellant grains to generate hot gas. It is stated: "Burning takes place by parallel layers where the surface of the grain regresses, layer by layer, normal to the surface at every point".

Propellant Burnout

- One of the issues to take care of, when choosing the burning rate of powder, is the time the powder takes to completely burn vs the time spent by the bullet inside the barrel.
- In the above ballistics graph, there is a change in the curve, at about 0.8 ms, at this point, the powder is completely burned, and new gas is not created.

Bore Diameter and Energy Transfer

- The maximum amount of energy is extracted by maximizing the swept volume.
- This can be achieved by any of the two ways – either the barrel length is increased, or by increasing the projectile diameter.
- If the barrel length is increased, the swept volume will increase linearly, while increasing the diameter will increase the swept volume as the square of the diameter.

The Ratio of Propellant to Projectile Mass

- Another factor, when choosing or developing a cartridge, is the recoil.
- The recoil is from the reaction of the projectile being launched, as well as from the gas of explosion, which exits the barrel with a velocity higher than that of the bullet.
- The powder-to-projectile-weight ratio also touches about efficiency.

Typical Propellant and Projectile Mass

For handgun cartridges, with heavy bullets and light powder charges (a 9 × 19 mm, for example, might use 5 grains (320 mg) of powder, and a 115 grain (7.5 g) bullet), the gas recoil is not a significant force; but for a rifle cartridge (a .22) Remington, using 40 grains (2.7 g) of powder and a 40 grain (2.7 g) bullet), the gas can be the majority of the recoil force.

Five General Equations Used in Interior Ballistics

- The equation of state of the propellant
- The equation of energy
- The equation of motion
- The burning rate equation
- The equation of the form function

Detailed discussion of the above equations is beyond the scope of this book. However, this has been mentioned for the sake of completeness of the topic and to draw the attention of the serious reader for further research on the subject.

3.2 EXTERNAL BALLISTICS

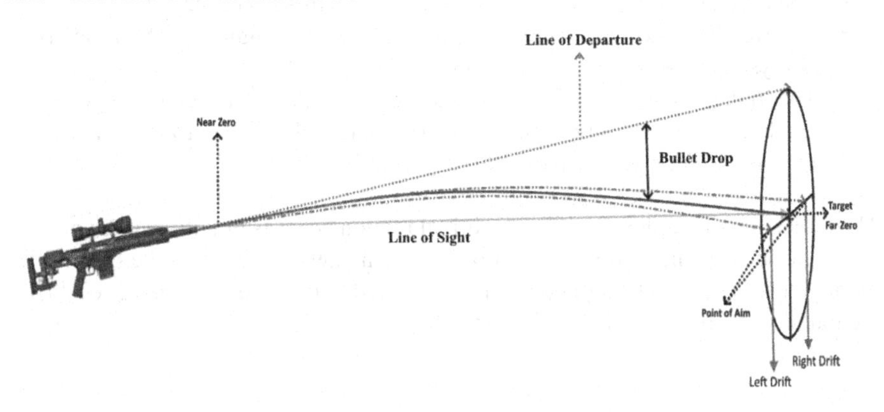

FIGURE 3.18 Exterior ballistic trajectories.

- Three forces influence a bullet's flight: Gravity, Air Resistance (Drag), and Wind.
- Gravity is a constant force that draws all objects toward the center of the earth. As soon as the bullet leaves the rifle muzzle it starts dropping toward the earth.
- So, if the flight time of the bullet is more, then the amount of vertical drop will be more because the bullet will be under the influence of gravity for a longer duration.
- The velocity of a bullet can minimize the effects of gravity. The faster a bullet can travel the distance to a target, the less time that gravity can act upon it.
- Since gravitational pull remains constant, it is possible to predict how much vertical drop a bullet will have, at a specific velocity and distance using ballistic charts.
- Air resistance or drag is the force that slows the bullet down. The amount of drag that a specific bullet experience varies depending on the bullet's velocity, shape, and weight.
- As the bullet travels, it must push air out of the way. This air is compressed in front of the bullet and then forced out of the way, sliding down the sides of the bullet.
- The movement of air down the sides of the bullet creates drag because of friction.
- The air moves from the sides of the bullet to the bullet's rear. This causes a vacuum to form, slowing the bullet even more. This drag can be reduced by selecting the proper shape and design for a bullet.

- Environmental factors of altitude, temperature, and humidity also create drag in that the denser the air, the bullet passes through the more drag it experiences.
- Bullets are manufactured in a variety of shapes, sizes, calibers, and materials.
- The efficiency of a bullet is referred to as the ballistic coefficient (BC).
- The BC (ballistic coefficient) is expressed as a three-digit number, which is always less than one. (e.g. .250, .530)
- The higher the number, the more efficient the bullet is in overcoming drag, the higher the ballistic coefficient, the flatter the bullet trajectory. See Figure 3.18 for Exterior ballistics Trajectories.

The bullet has a complex motion. The initial motion consists of nutation and precession. The subsequent motions consist of continual yaw coupled with the bullet spin. Finally, when the bullet slows down and is allowed to proceed without any obstacles the bullet starts tumbling before landing on the earth (Figure 3.19).

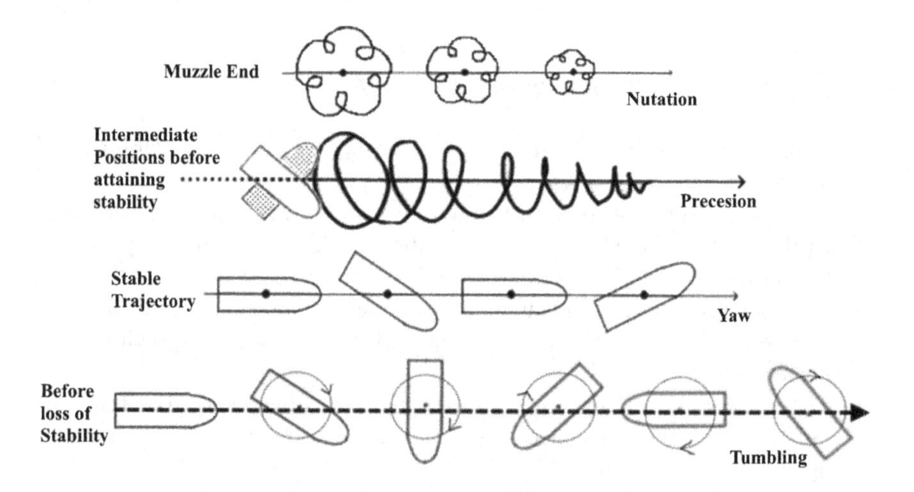

FIGURE 3.19 Transition phase motion of bullet from muzzle end.

Soon after the projectile's motion of nutation and precession dies down, its trajectory is governed by the following forces:

1. **The force of gravity (g)** – It is the force that acts to the center of the projectile.
2. **Aerodynamic forces** – It is the force that acts through the center of the pressure depended on the shape of the projectile. The aerodynamic forces consist of three types of drag i.e. Skin Drag, Shape Drag, and Shockwave Drag which act in the direction opposite to motion. Also, there is a force of lift that acts perpendicular to the direction of the motion.

The forces are given by the following formulas

Drag forces:

$$F_{drag} = C_d \times \frac{1}{2}\rho v^2 A \qquad \text{Eq 3.3}$$

C_d = Coefficient of the drag.
A = Effective area of the projectile body.
Now,

$$C_d = \frac{\text{Drag Force}(F_{drag})}{\text{Dynamic Force}},$$

$$BC = \frac{\text{Sectional Density}}{\text{Coefficient of Drag}} = \frac{\left(\frac{m}{A}\right)}{C_d} = \frac{m}{C_d \times A}, \qquad \text{Eq 3.4}$$

Using Equations (3.3) and (3.4), we can also write:

$$F_{drag} = \frac{1}{2} \times \frac{1}{BC} \times \rho v^2 m \qquad \text{Eq 3.5}$$

where ρ is the density of medium and v is the velocity of the projectile.

Where m is the mass of the projectile and BC is the ballistic coefficient which is not a constant and function of Mach number. At the transonic zone (0.8–1.2 Mach) the ballistic coefficient undergoes an abrupt change. Due care to be exercised to evaluate the deceleration (r) in this zone.

$$r = \frac{F_{drag}(\text{drag force})}{m} = \frac{1}{2} \times \frac{1}{BC}\rho v^2 \qquad \text{Eq 3.6}$$

The retardation of the bullet is inversely proportional to the ballistic coefficient.

$$\text{Lift Force: } F_{Lift} = C_L \frac{1}{2}\rho v^2$$

C_L= Coefficient of lift that depends on the angle of attack.

The aerodynamic stability is dependent on the location of the center of pressure with respect to the center of mass of the projectile. If the center of pressure is behind the center of mass when the body changes the angle of attack, a restoring force is created to bring the body back in the neutral direction and therefore the projectile remains stable. The stability is further improved by the spin. Any change in direction of the linear motion brings into action, the gyroscopic torque which resists change in orientation of the projectile in motion. The schematic of the flight stability is illustrated in Figure 3.20.

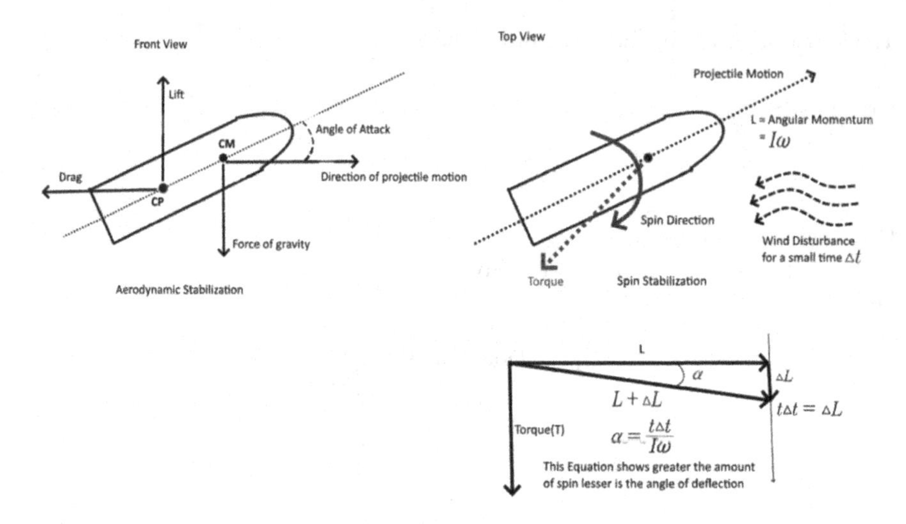

FIGURE 3.20 The schematic of flight stability.

1. **Coriolis force** – The Coriolis effect causes a drift in a direction perpendicular to the axis of the earth's rotation. Its horizontal component causes a curvature at the trajectory in the right direction on the horizontal plane in the northern hemisphere and the left direction in the southern hemisphere. The horizontal drift is the highest at the pole and zero at the equator. The drift in the vertical curvature is largest at the equator and decreases to zero at the pole. $F_{coriolis} = 2\ v\ \Omega\ m$, where m = mass of projectile, v = velocity of projectile and Ω is the rotational speed to earth (3.5×10^{-5} rad/s). This force becomes important if the projectile travels a longer distance. For example if the average projectile speed is 1,000 m/s, the acceleration is $2.0 \times 1{,}000 \times 3.5 \times 10^{-5} = 0.07$ m s^{-2} and in a direction perpendicular to the velocity.

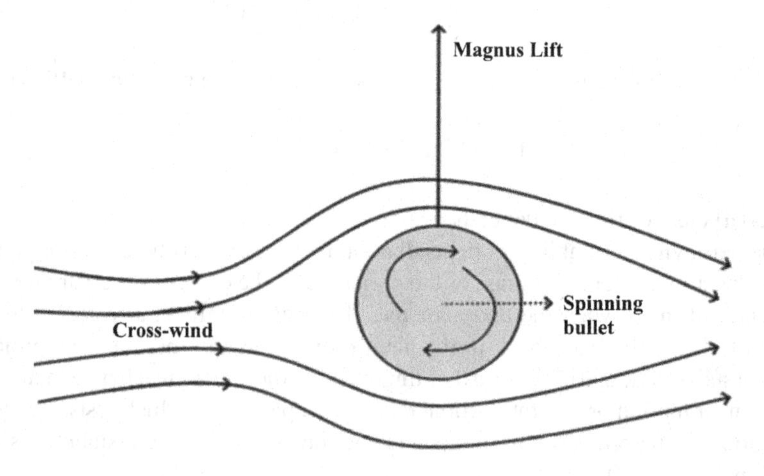

FIGURE 3.21 Drift force.

2. **Drift force** – It is a destabilizing force on account of the projectile spin. When the bullet follows the parabolic path, the curvature of the path simulates a precession motion of the bullet about the instantaneous center of the curvature of the trajectory. And this results in a drifting force perpendicular to the direction of the motion. Depending on the direction of spin, the bullet will either drift in the right or left of the point of aim.
3. **Magnus effect** – It is created by the spin of the bullet that creates an up or downward force perpendicular to the direction of the wind vector, depending on the direction of the spin of the bullet. The Magnus force is significant only when wind speed exceeds 4 m/s. The effect of the Magnus force is to change the location of CP (center of pressure) with respect to the CM (center of mass), if it shifts the CP behind the CM, it has a stabilizing effect on the bullet. But if CP goes ahead of CM, it destabilizes the projectile (Figure 3.21).

3.3 TERMINAL BALLISTICS

The study of the effect of the projectile upon its reaching the target is called terminal ballistics. At the terminal, the projectile is expected to perform some specific functions as preplanned. The projectile can simply hit the target, and transfer some desired amount of energy just to penetrate it. The projectile may carry some incendiary or explosive head which may burst upon hitting the target, or it may be so designed that after reaching the designated target it may explode in the air over the target, and cause planned damage. Therefore, the design of the projectile is very much important from point of view of its effects on the target being shot at. The terminal ballistics, therefore, needs special attention, and itself is a sub-field of the ammunition ballistics.

FACTORS AFFECTING THE AMOUNT OF DAMAGE DONE BY THE PROJECTILE

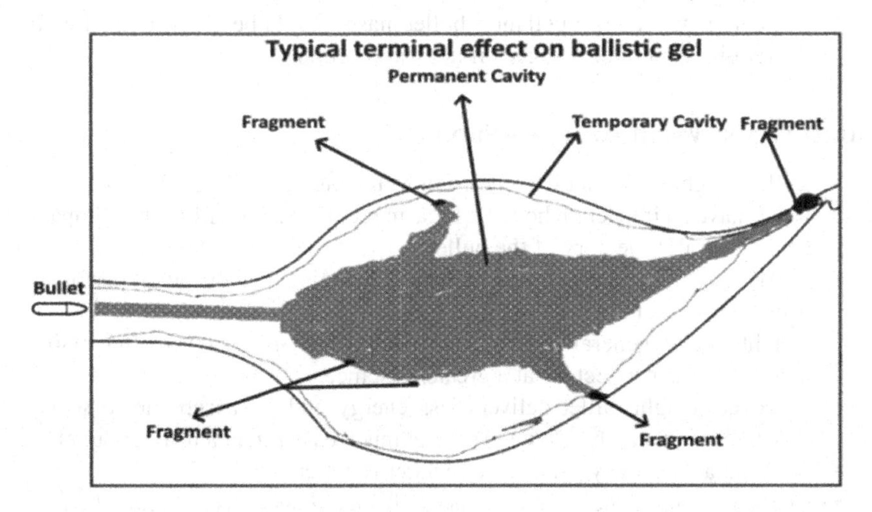

FIGURE 3.22 Terminal effect on ballistic gel.

- Bullet Mass **Figure 3.22**
- Bullet Velocity
- Air Drag
- Bullet Size
- Bullet Geometry
- Bullet Deviation
- Cavitation Factor

The bullet causes damage to the target by laceration and crushing along its track through the body. The size of the crush injury in the tissue is the same as the order of the diameter of the bullet.

By a high-speed bullet, a permanent cavity is made by the path of the bullet itself by crushing the tissue. A temporary cavity may also be formed by the projectile due to the stretching of the wound in the process of energy absorption. This effect is known as the Cavitation effect.

The bullet creates shockwaves that travel ahead of the bullet in the medium, also spread energy sideways. The high pressure generated which can reach up to 200 times the atmospheric pressure, when moving at high velocity can create profound injury. The pressure so generated even from a distant bullet can induce neurological disorders in humans.

3.4 HANDGUN AMMUNITION

Desired Characteristics of Handgun Ammunition

- They should have enough power and accuracy to hit the target at a distance not exceeding 50 m.
- It should have weight, shape, muzzle velocity, and energy consistent with manageable recoil.
- The cartridge load (propellant + bullet mass) should be synchronized with the requirement of creating semi-auto/auto action.

Bullet Grain Weight Impact on Shooting

- Bullet weight makes a difference in how the bullet performs.
- It will have an impact on how the firearm recoils, and it will have an impact on the overall trajectory of the bullet.
- Finally, it will impact terminal ballistics, changing how the bullet performs once it strikes a target.
- A lighter bullet generally has greater speed, which means it can often maintain a straighter trajectory at a greater distance.
- However, a light bullet delivers less energy and is susceptible to wind, which means it can be blown off target much easier than a heavier bullet.
- So, from an accuracy perspective neither is better.
- Check for sample idea of recoil energy(ft-lb) in handguns in Figure 3.23.

Bullet Weight and Recoil

- According to the laws of physics, bullet grain plays an important role in recoil. It requires more energy to move a heavier bullet than a lighter bullet.
- However, the recoil is the product of bullet mass and velocity and recoil energy are in the proportion of the ratio of bullet mass and the weapon mass, hence heavier bullet means more recoil.
- It is directly proportional to the bullet grain and the square of the velocity and inversely proportional to the weapon weight.
- In a handgun weighing from 1 lb to 3 lb, the recoil energy is in the range of 2–9 ft-lb. Some examples are given below (Table 3.1).

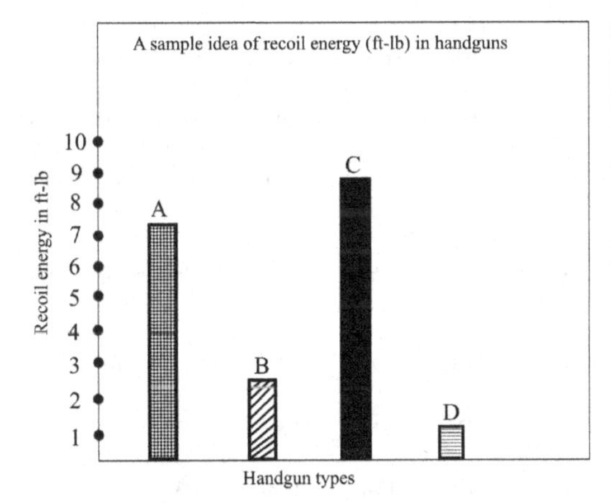

A	9 x 19mm +P 115gr at 1250 fps/ Handgun weight -> 1.5 lbs
B	0.380 ACP/ 90gr at 1000fps/ Handgun weight -> 1.5 lbs
C	0.375 Magnum/ 125gr @ 1209 fps Handgun weight -> 1.75 lbs
D	0.32 S&W/ 100gr @ 700 fps/ Handgun weight -> 2.0 lbs

FIGURE 3.23 A sample idea of recoil energy (ft-lb) in handguns.

A chaotic population of information is available on the handgun ammunition, bizarre enough to confuse even the experts in the field. So, simple metrics to select the weapon have been presented after analysis of the ballistics.

The stopping power is the main criteria for the personal defense weapon. The terminal ballistics is the determiner of the “man-stopper”. The bullet shape, weight, and velocity at the target decide the terminal effect. It’s the remaining K.E. at the target that is important for the projectile.

It’s a well-known fact that a minimum of 60 ft-lb of energy is required to be transferred to the vital human organ to stop. The handgun and ammunition are to be chosen accordingly. The sectional density of the projectile is directly proportional to the ballistic coefficient, hence the remaining energy. It may be easily understood that for stopping the target higher PSD ammunition and longer barrel weapon will be necessary.

High bullet weight and higher MV means high recoil. This is also to be matched with the individual threshold limit of withstanding recoil. Bullet stability is another factor to be considered in weapon and ammunition selection. The spin of the projectile

TABLE 3.1
Wade through the Handgun Ballistics

Barrel length (inch)	Bullet Type	Bullet cal	Diameter (mm)	Bullet Weight (gr)	Muzzle Velocity (fps)	V 50 (yds)	V 100 (yds)	KE (ft-lb) Muzzle	KE ft-lb 50	KE ft-lb 100 yds
2	Round Nose	0.25 Auto	6.35	50	760	707	659	64	56	48
	Slug	0.25 Auto	6.35	45	815	729	655	66	53	42
	Slug	0.25	6.35	35	1150	-	-	103	-	-
3	Round Nose	0.32 S&W	8.12	85	680	645	610	90	81	73
	Slug	0.32 Auto	8.12	55	1200	-	-	176		
4	Round nose	0.38 S&W	9.65	145	685	650	620	150	135	125
		0.38 SPL	9.65	158	755	723	693	200	183	105
	Wad Cutter	0.38 SPL Wad	9.65	158	755	721	689	200	182	167
		0.38 SPL P Wad	9.65	158	890	855	823	278	257	238
	Hollow Point	0.38 SPL P	9.65	110	995	926	871	242	210	185
	Jacketed	9mm Luger	9	115	1225	1095	1007	383	306	259
		0.357 Magnum	9.06	125	1450	1240	1090	583	427	330
5	Slug	0.357 SIG	9.06	80	1650	-	-	484	-	-
		10mm Auto	10	115	1650	-	-	695	-	-
		400- Con Bor	10.16	115	1650	-	-	695	-	-
	Full Metal	0.45 Auto	11.43	185	770	707	650	244	205	174
	Jacketed	0.45	11.43	230	835	800	767	365	326	300
	Hollow Point	0.45 Auto	11.43	185	1000	938	888	411	362	324
	Jacketed	0.45 GAP	11.43	185	1090	970	890	490	385	325

is an important factor to decide the stability of the bullet. Greens and Miller's formula for determining the twist for the required stability is well known. However, as a rule of thumb, the sub-sonic ammunition with the lead core is fired through the barrels with a twist of 1:18 to 1:48 and the exact parameter is to be selected based on the required range and remaining K.E.

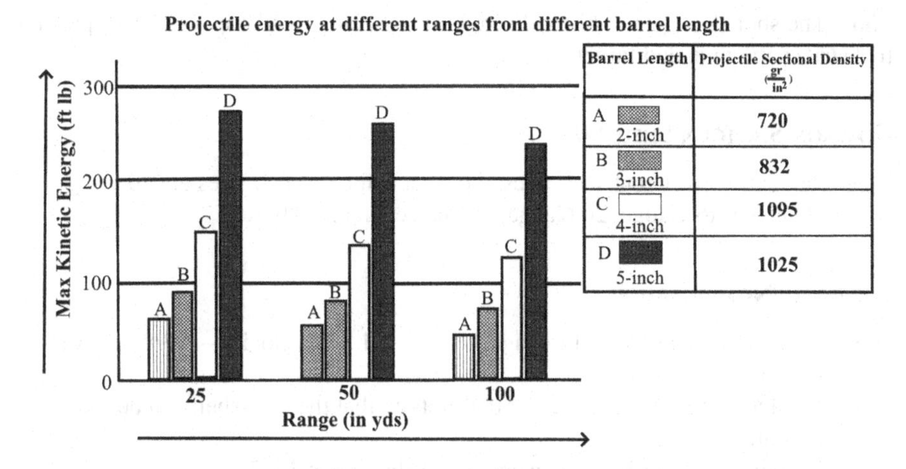

FIGURE 3.24 Projectile energy at different ranges from different barrel length.

3.5 SHOTGUN AMMUNITION (FIGURE 3.25)

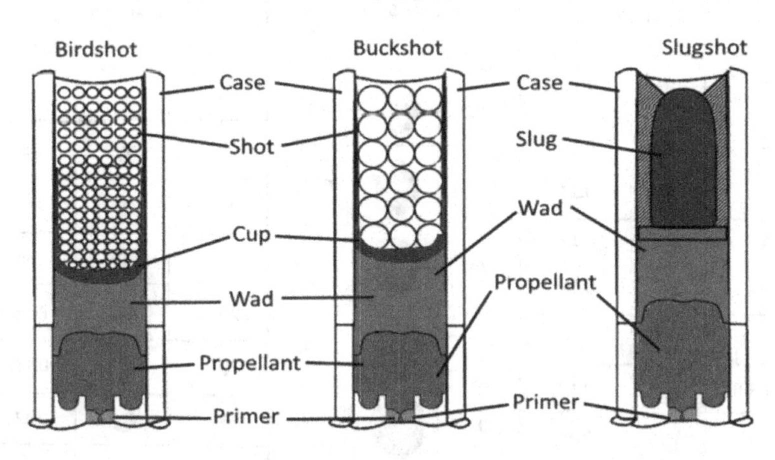

FIGURE 3.25 Shotgun shells – shotshells main parts.

Hull: The hull contains the primer, powder, wad, and shot pellets.

Primer: The primer's internal compound explodes when struck by the firing pin and ignites the powder.

Gunpowder: The powder burns and creates expanding gases to move the wad, shot cup, and shot down the bore.

Wad: The gas behind the shot charge is sealed by the wad.

Shot Cup: The shot cup protects the shot and the internal portion of the barrel.

Shot: The shot (lead, steel, bismuth, iron, nickel-plated, and copper-plated, pellets) that strikes the intended target.

Common Shotgun Shell Sizes

- Shotgun has different shell sizes: the most common shell sizes are 10 Gauge, 12 Gauge, 16 Gauge, 20 Gauge, 28 Gauge, and .410 bore.

Shotshell Nomenclature

- Standard specification: 12 gauge – 3 1/2" – 1,550 Velocity – 1 3/8 oz – BB Shot
- 12 Gauge refers to the gauge of the shotgun that the shot-shell can be safely fired in.
- The chamber length of the gauge of the shotgun is 3 1/2".
- The total weight of all the shot pellets loaded in the shell is 1 3/8 oz.
- #BB shot is the shot pellet size that is loaded in the shotshell. See Figure 3.26 for more sizes.

STEEL SHOT		
Number		SIZE Inches
6		0.11
5		0.12
4		0.13
3		0.14
2		0.15
1		0.16
Air Rifle		0.177
BB		0.18
BBB		0.19
T		0.20
F		0.22

Buck Shot		
Number		Size Inches
4		0.24
3		0.25
2		0.27
1		0.30
0		0.32
00		0.33
000		0.36

Lead Shot		
Number		Size Inches
12		0.05
9		0.80
8-1/2		0.85
8		0.90
7-1/2		0.95
6		0.110
5		0.120
4		0.130
2		0.150
BB		0.180

Sizes based on American standard.

FIGURE 3.26 Shot sizes.

Shot Ballistics

- Maximum Effective Lethal Range, Retained Pellet Energy, and Penetration.
- Regardless of the type of shot used, whether it's lead, steel, bismuth, or tungsten, there must be the following minimum pellet penetration depth in inches to cleanly kill the following targets.
- 1.5 inches of penetration for large ducks, a minimum of 1.75 inches of penetration for pheasants, and a minimum of 2.25 inches of penetration for large geese at the ranges decided to take them.

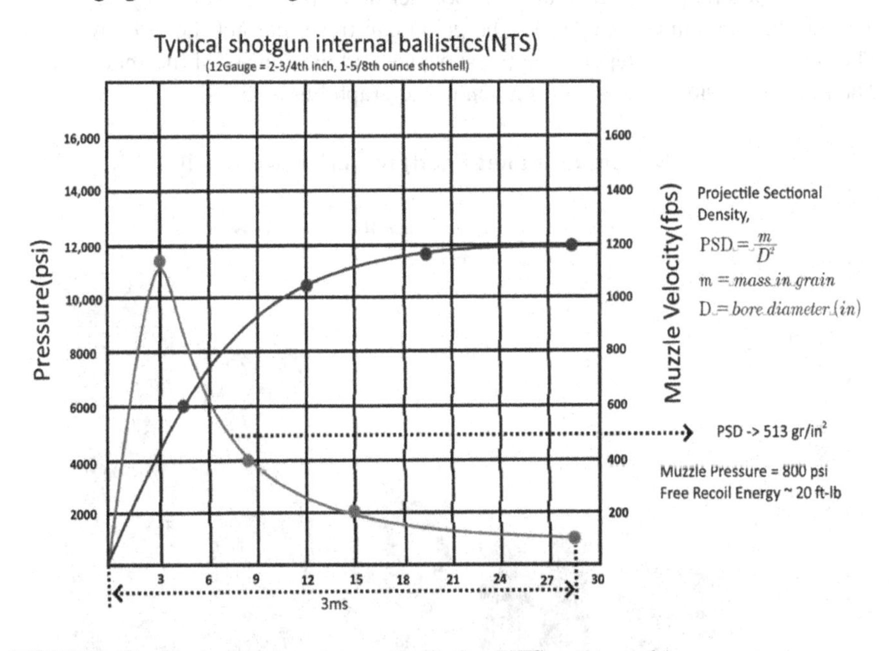

FIGURE 3.27 Typical shotgun internal ballistics (NTS- not to scale).

It may be noted that for a self-loading, semi-automatic shotgun, the gas port for automatic action cannot be further away from the chamber as there won't be enough gas pressure for the creation of automation. Accordingly, a semi-auto shotgun has a gas port immediately after the chamber to avail the necessary pressure. The above graph shows the internal ballistics of a shotgun using slug or sabot which is very similar to that of a rifle and pistol (Figure 3.27).

Exterior Shot Ballistics

After exiting the barrel, the volume of the shots expands both in length and width and creates a shot column. The shape of the forcing cone from the chamber to smoothbore, the choke, and the size of the shot determine the range and spread of the shot column. The shot projectiles are spheres with a poor ballistic coefficient.

It is difficult even to characterize the path of a single sphere of given mass moving through the air with a given initial velocity. In the shot column, there are interactions of several shots traveling together, and the situation makes it difficult to predict the path of moving shot columns, and it is a shot-column, not a single projectile that travels to the target. The success of the hit depends on the interception of the target with shot-column. So, the steering or aiming of the gun while shooting plays an important role.

A shotgun range is a function of pellet velocity and its mass, in general #9 shots as a maximum range of 175 yd and has 00 buckshot has a range of 740 yd. However, these are the maximum range that the pellet can travel but not the effective range. The range of shotguns depends on the shot size and the density of the shot material. The range of various lead shots is given in the graph Figure 3.28.

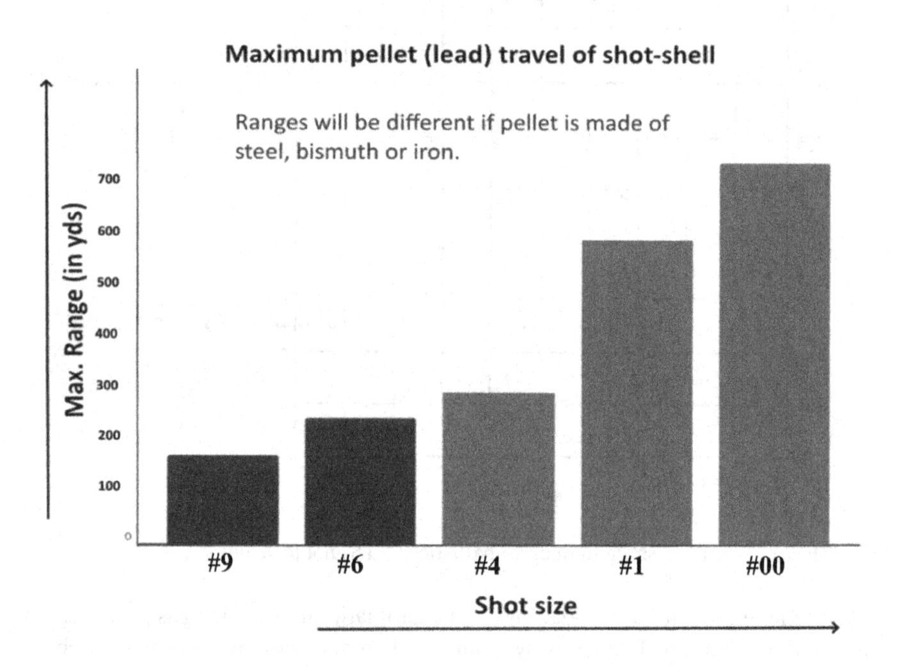

FIGURE 3.28 Maximum pellet(lead) travel of shot-shell.

Shotshell Application

Shotshells (with pellet sizes ranging from #12 to #000) are fired for different applications.

1. Skeet shooting – This is to hit a fast-moving target at close ranges; so many small pellets are required for the sport. The typical shot size is #9. The range of skeet shooting is 25–30 yd, the objective is to maximize the number of pellets in the close range. So, a wide choke shotgun is chosen for this purpose.

2. Trapshooting – Trapshooting is the game when the shooter engages a target moving fast away from him, requires the pellets to have enough momentum to catch up with the fleeing targets. Therefore, larger pellets of #8 or less will be required to shoot the target. However, the sports demand that minimum damage be caused to the target, so pellet size not exceeding #7.5 is used. The range of trapshooting is 45–60 yards. So, a more focused shot column is wanted. Hence the weapon of choice must have a tighter choke.

   The degree of choke is varied from wide to tight and covered by the range from cylinder to extra-full. To have a variable choke, the barrel is designed with different screwed-on chokes.
3. Buck shooting – The target requires enough energy to take it down, so larger buck shots and slugs are used for this purpose.
4. Self-defense – Enough momentum is required to stop the offender, so shotshells #3 and #4 are used. They have enough energy to make the enemy immobile but not kill him within a range of 10–20 yards.

BIBLIOGRAPHY

G. Burrard, *The ID of Firearms Forensic Ballistics*, Herbert Jenkins Ltd., London, UK, 1934.

G. Frost, *Ammunition Making*, The National Rifle Association of America, Washington DC, USA, 1990.

P. Labbett, *Military Small Arms Ammunition of the World, 1945–1980*, Presidio Press, San Rafael, California, USA, 1980.

Richard A. Mann, and Jim Wilson, *Handgun Training for Personal Protection: How to Choose Use the Best Sights, Lights, Lasers Ammunition*, Gun Digest Books and Krause Publications, Iola, Wisconsin, USA, 2013.

4 Anatomy of Small Arms

THE PRINCIPLE OF DESIGN OF ANATOMY

The configuration and anatomy of the small arms can be better appreciated as machines for offense or defense by viewing them from the combined platform of IC engines and vented fired pressure vessels to ensure complete burnout.

The engine platform decides the body, barrel, firing, and feed mechanism. This also decides the gas flow path and propellant energy distribution and muzzle devices. The vented pressure vessel aspects determine the safe operating pressure of the weapon and the choices of materials for the major force-bearing parts.

The choice of operating principle is a matter of reliability and maintenance of cleanliness inside the housing body and trigger mechanism. This also affects the complexity and the costs of the weapons.

The safety requirements are provided by incorporating "AND" logic in the trigger/firing mechanism.

Attachments are provided to fix add-on accessories on the weapons.

The feed mechanism is decided based on the role of the weapon and demand for the volume of fire.

The sight opts for the accommodation of firer reflexes and target position and the intended hit probability.

The challenge of designing the most efficient and easily maneuverable weapon is to ensure complete propellant burnout balancing muzzle blast, muzzle thrust, and ensuring the least perceived recoil.

The thermodynamics also poses a challenge of overheating the weapons designed for a high volume of sustained fire. Therefore, the design of the operating gas flow path is of paramount importance for efficient weapon design.

The ergonomics and aesthetics play a vital role in the confidence of the user. Therefore, the selection of form, materials, and surface treatment of the exterior plays a vital role to enhance the acceptability of the weapon system.

Finally, any firearms work with the basic cyclic operations of feed–load–fire–extract–eject sequence. So, the weapon architecture must be evolved in the simplest possible manner to reliably satisfy these elementary functions.

Last but not least and exhaustive facts about firearms are that these are not merely a kinematic synthesis of various metal parts, as it seems to be assumed even in the weapon systems of repute. The inertia of the moving parts and the frame as well play important roles in the reliable functioning of the weapon. Thus, constituent materials and their properties, in addition to the size and shape of components, are of paramount importance to ensure the reliability of the automatic firearms.

DOI: 10.1201/9781003199397-4

TABLE 4.1
AK 47**

General Specification	
Type	Assault Rifle
Place of Origin	Soviet Union
Designer	Mikhail Kalashnikov
Weight	6.83 lb. without magazine
Length	35" Fixed Wooden Stock
Barrel Length	16.3"
Cartridge	7.62 × 39 mm
Action	Long Stroke Gas, rotating closed bolt.
Rate of Fire	600 RPM
Muzzle Velocity	2350 ft/s
Effective Range	440 yds
Feed System	30-rounds detachable box magazine 100-rounds drum magazine
Sight	Iron sight

FIGURE 4.1 AK-47.

AK 47

AK 47 is illustrated here for its classic architecture in the design of gas flow paths. The weapon is surprisingly reliable in firing underwater. It is queer to notice that the weapon fires at a much faster rate under the water! Given a Super Cavitation bullet, the weapon will easily lean itself to a design for an underwater firearm. So, the weapon is of utmost importance from a serious researchers' point of view (Table 4.1 and Figures 4.1–4.3).

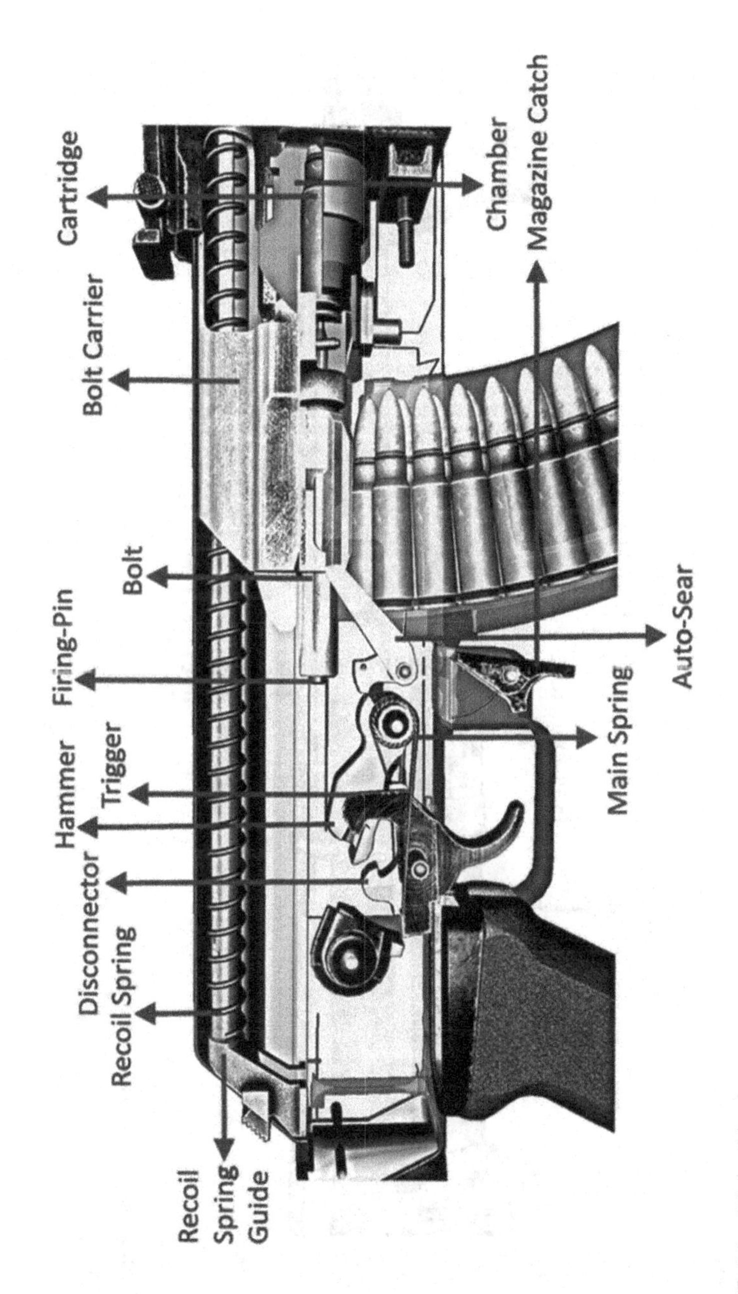

FIGURE 4.2 Right side view.

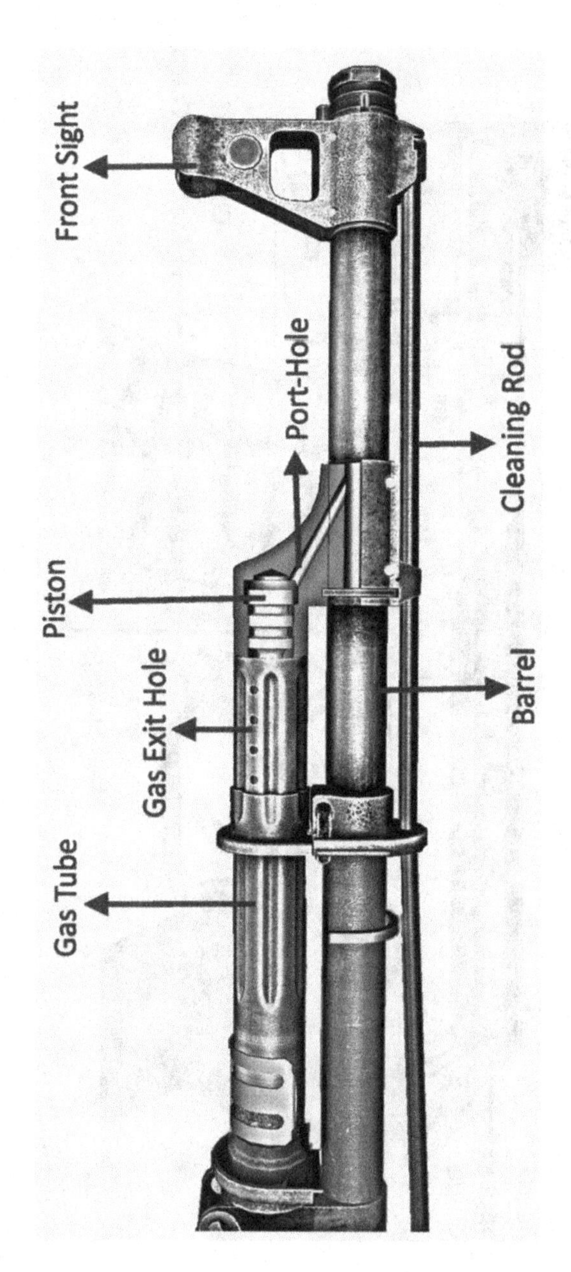

FIGURE 4.3 Right side barrel view.

Browning HP

(Table 4.2 and Figures 4.4 and 4.5)

TABLE 4.2
Browning HP**

General Specification

Type	Pistol Semi-Auto
Place of Origin	Belgium
In-Service	From 1935
Designers	John Browning, Dieudonné Saive
Weight	2.19 lb
Length	7.8"
Barrel Length	4.7"
Cartridge	7.65 × 21 mm Parabellum 0.40 SW 9 × 19 mm Parabellum
Action	SA (Single Action), Short Recoil
Muzzle Velocity	1100 ft/s using 9 mm parabellum cartridge
Effective Range	54.7 yds
Feed System	10-rounds, 13-rounds detachable box magazine
Sight	Stock dot sight.

FIGURE 4.4 Browning HP.

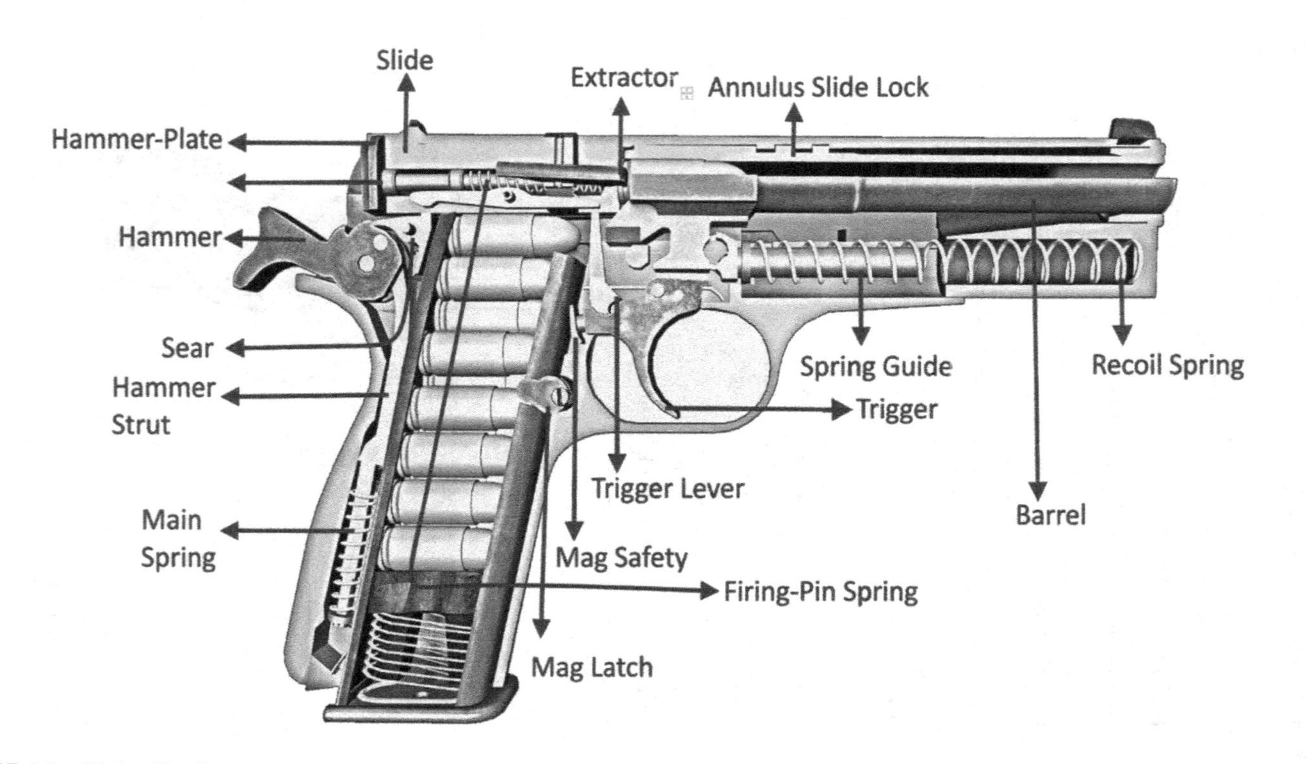

FIGURE 4.5 Right side view.

S&W 53 Revolver

(Table 4.3 and Figures 4.6 and 4.7)

TABLE 4.3
S&W 53 Revolver**

General Specification	
Type	Revolver
Place of Origin	USA
Designers	Smith & Wesson
Weight	2.5 lb.
Length	11¼ "
Barrel Length	6"
Caliber	0.22 LR 0.22 Remington Jet
Action	Double Action Only (DAO)
Feed System	6 rounds cylinder
Sights	Fixed front post and rear notch

FIGURE 4.6 S&W model 53.

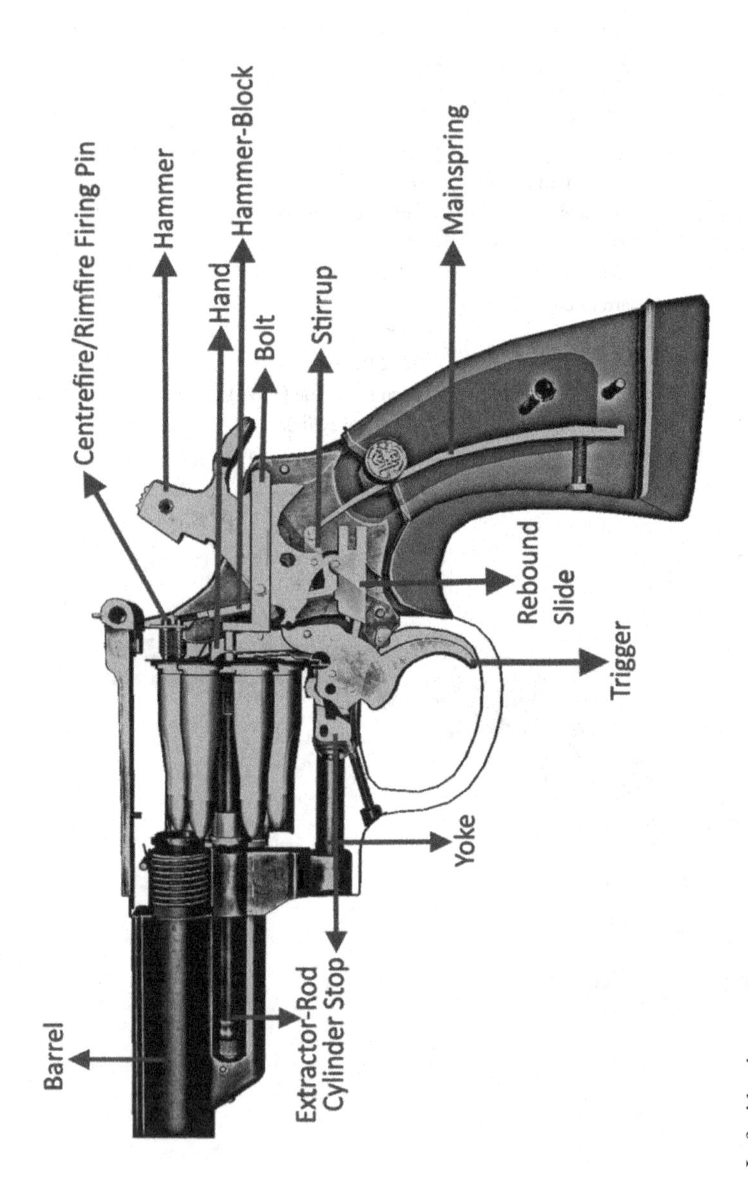

FIGURE 4.7 Left side view.

ShotGun: Remington 870

(Table 4.4 and Figures 4.8–4.10)

TABLE 4.4
Shot Gun: Remington 870**

General Specification

Type	Shotgun
Place of Origin	USA
Designed in	1951
Designers	L. Ray Crittendon, Philip Haskell, Ellis Hailston, G.E. Pinckney
Weight	7.0 lb
Length	37.25"
Barrel Length	18"
Cartridge	12-, 16-, 20- , 28-gauge or 0.410 bore
Action	Pump action
Feed System	Round internal tube magazine capacity (4+1) or (7+1)
Sight	Adjustable Iron Sights with cantilever and receiver-mounts for scopes

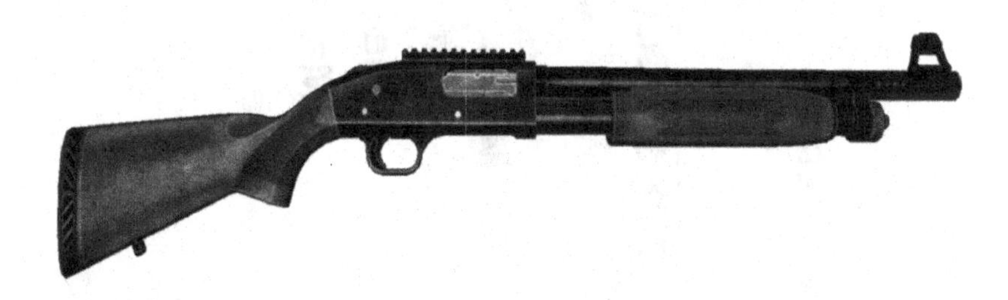

FIGURE 4.8 Remington 870.

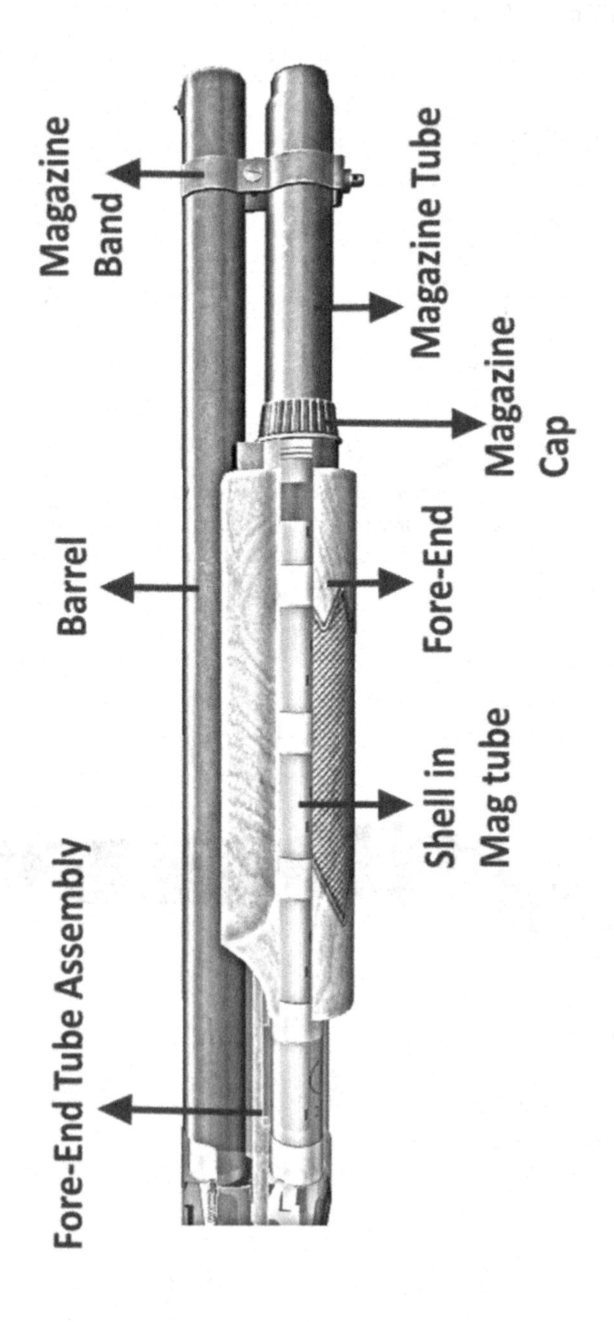

FIGURE 4.9 Right side barrel view.

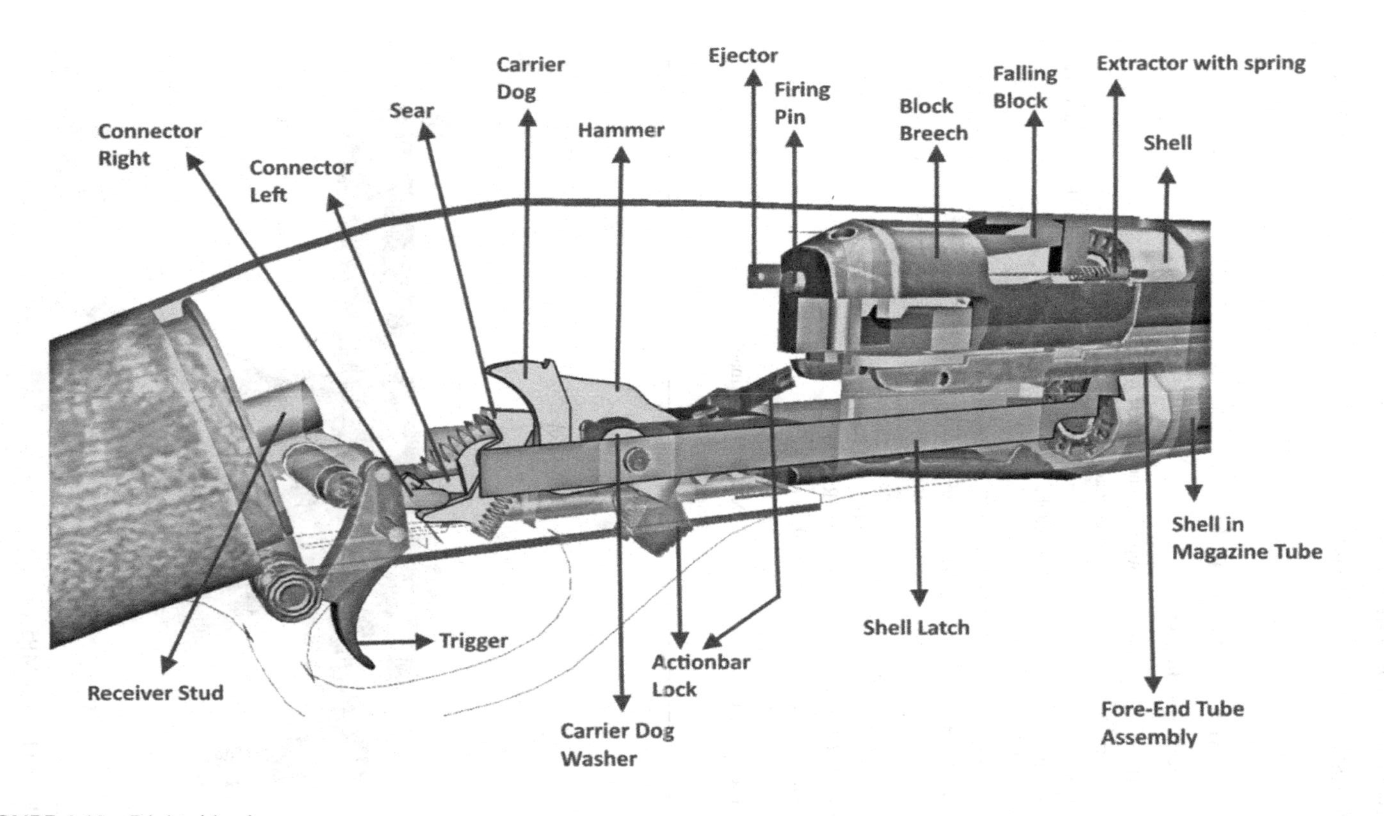

FIGURE 4.10 Right side view.

Sub-Machine Gun HK MP5

(Table 4.5 and Figures 4.11 and 4.12)

TABLE 4.5
Sub-Machine Gun- HK MP5**

General Specification	
Type	Sub-Machine gun
Place of Origin	West Germany
Designers	Tilo Moller, Manfred Guhring, Georg Seild, Helmut Baureuter
Weight	5.5–6.8 lb.
Length	14.5–31.1"
Barrel Length	4.5–8.9"
Cartridge	9 × 19 mm Parabellum 10 mm Auto 0.40 S&W
Action	Roller Delayed Blowback closed bolt
Rate of Fire	800–900 RPM
Muzzle Velocity	935–1,394 ft/s
Effective Range	230–636 ft
Feed System	15–40 Detachable box mag
Sight	Iron, Rear rotary drum, front hooded post

FIGURE 4.11 HK-MP5.

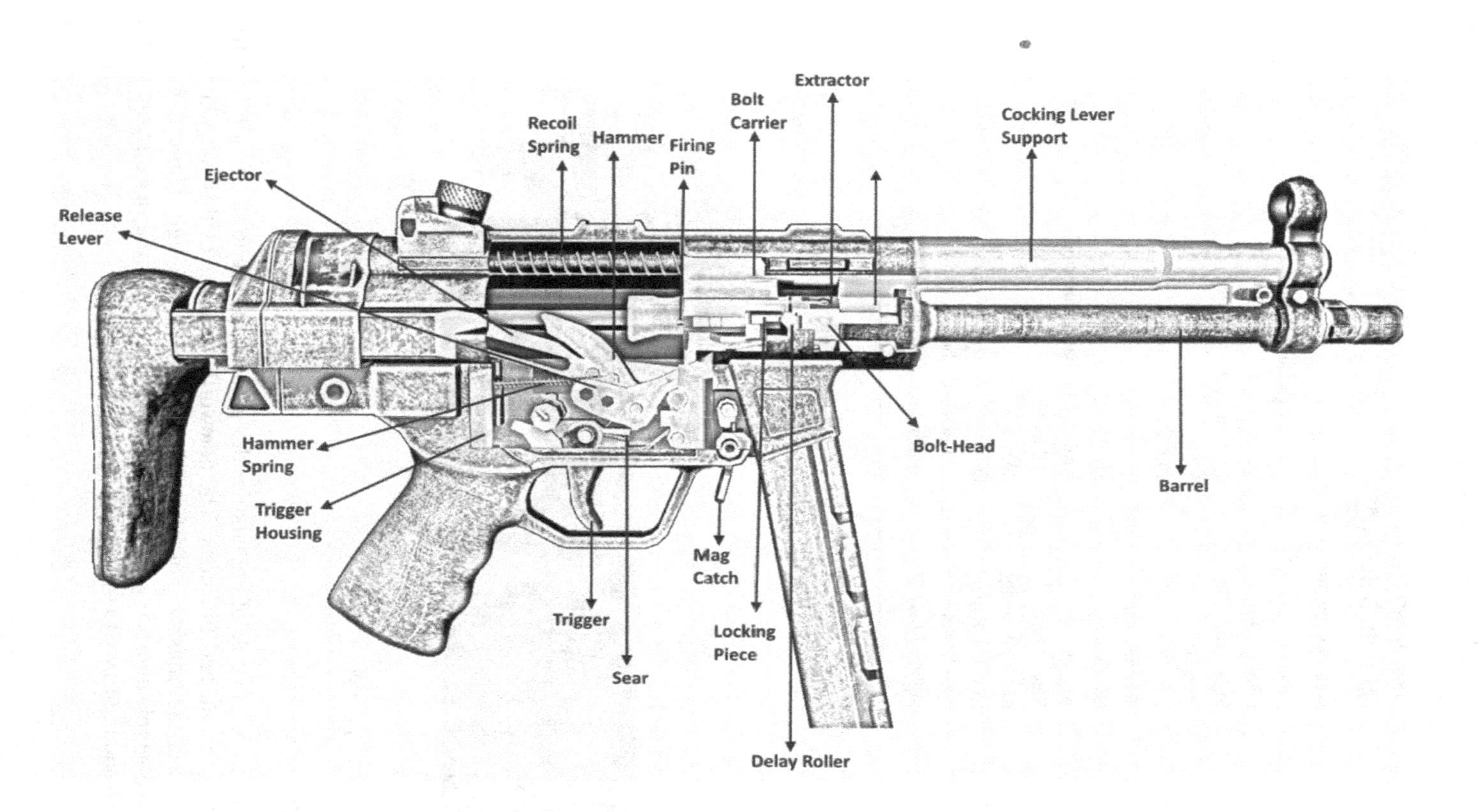

FIGURE 4.12 Right side view.

TABLE 4.6
FN P90, Belgium**

General Specification	
Type	Personal Defence Weapon (PDW)
Place of Origin	Belgium
In-Service	1991–present
Weight	5.7 lb.
Length	19.9"
Barrel Length	10.4"
Cartridge	FN 5.8 × 28 mm
Action	Straight Blowback, closed bolt
Rate of Fire	900 RPM
Muzzle Velocity	2350 ft/s
Effective Range	660 ft
Feed System	50 Detachable box magazine
Sight	Iron and Reflex sight

FIGURE 4.13 FN-P90.

FN P90

This is a classic example of ergonomics design using plastic to satisfy all the attributes of forms, fit, and functions. The noteworthy features of this weapon are its paraxial magazine, consisting of 50 rounds or so to deliver a high volume of fire in a closed combat role in urban warfare as often being seen to occur in a terrorist attack (Table 4.6 and Figures 4.13–4.15).

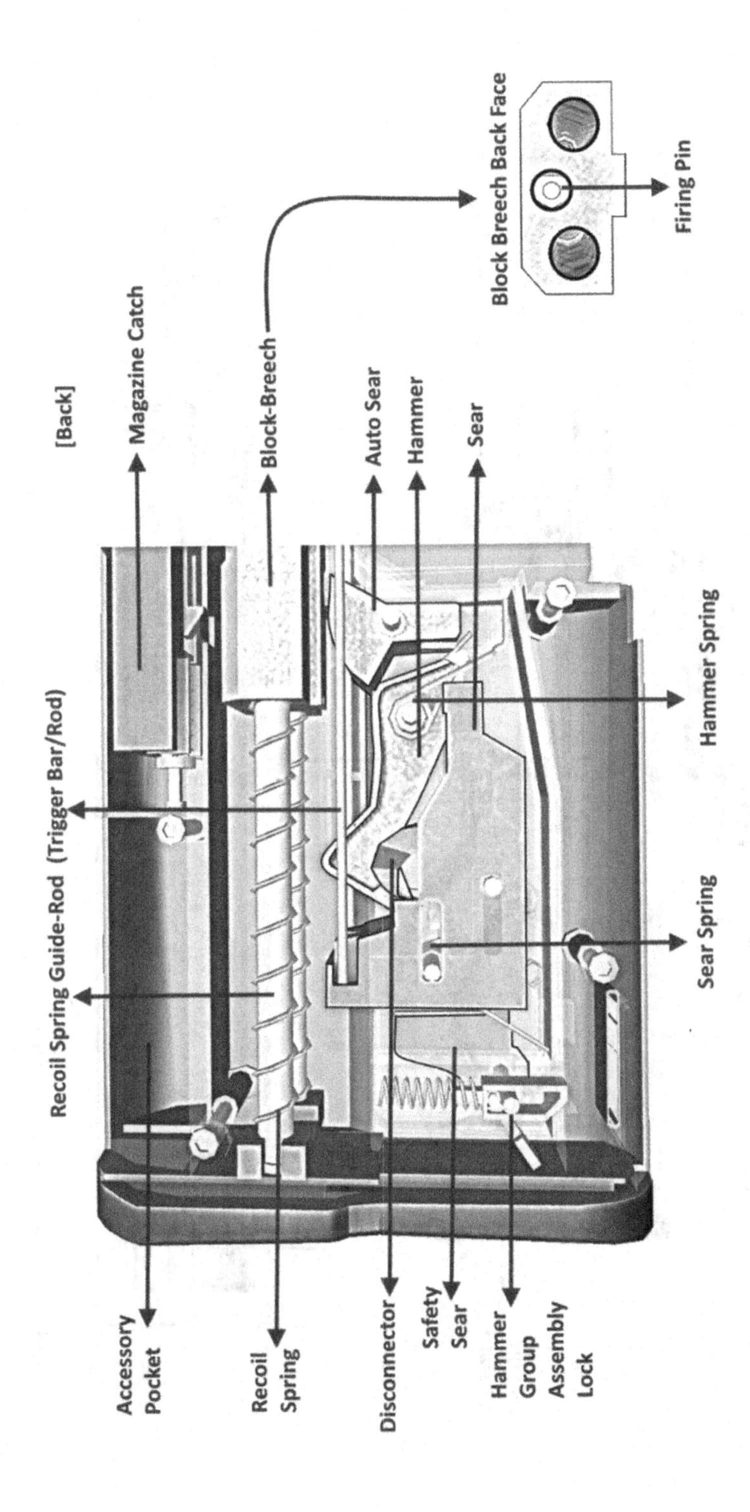

FIGURE 4.14 Right side receiver view.

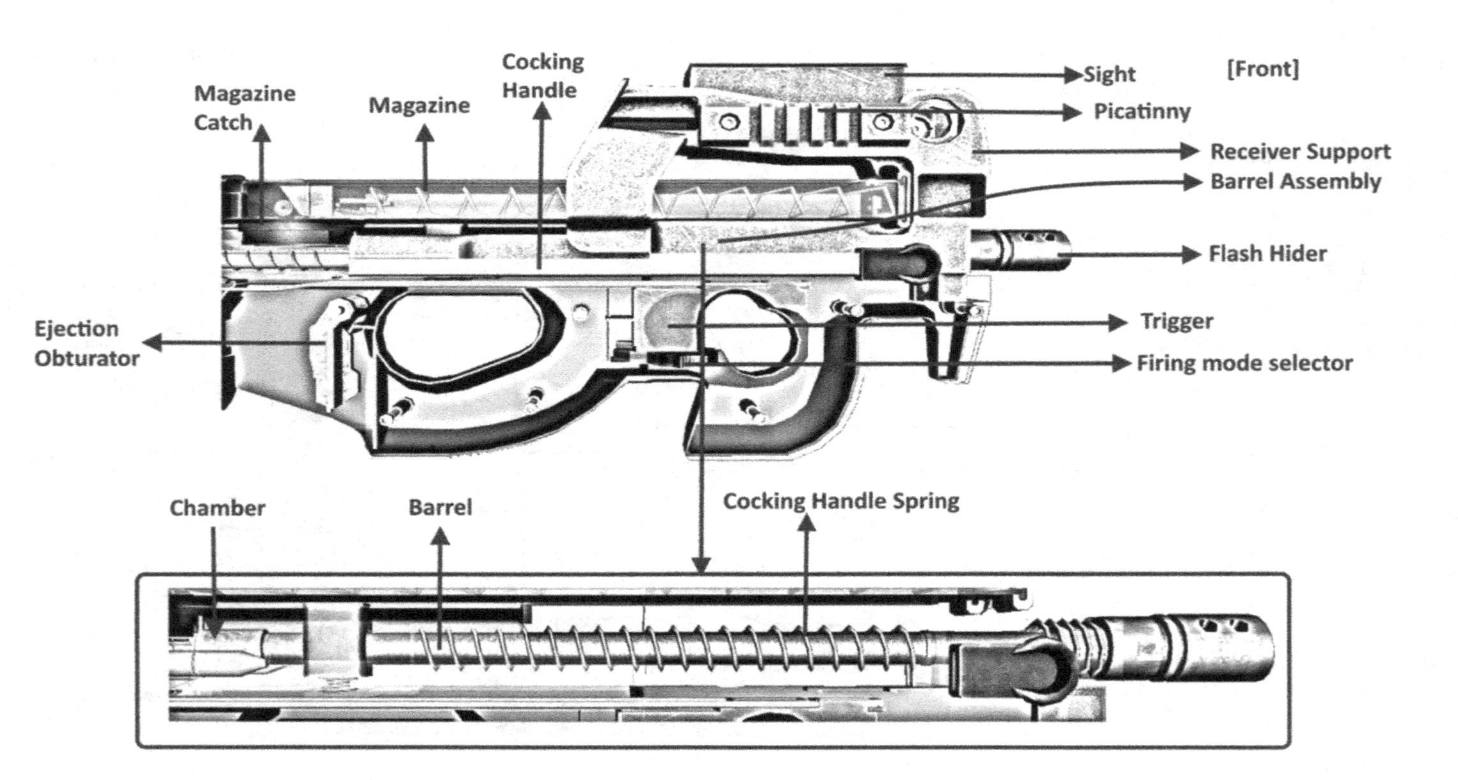

FIGURE 4.15 Right side view with barrel assembly.

TABLE 4.7
AN 94**

General Specification	
Type	Assault Rifle
Place of Origin	Soviet Union, Russia
Designer	Gennadiy Nikonov
Weight	8.49 lb.
Length	28.7–37.1"
Barrel Length	15.9"
Cartridge	5.45 × 39 mm
Action	Gas-operated (Recoil Combo), rotating closed bolt.
Rate of Fire	1,800 RPM (2rd Burst)
	600 RPM (Full Auto)
Muzzle Velocity	2,953 ft/s
Effective Range	2,296 ft
Feed System	30-, 45-rounds AK 74 compatible detachable box magazine
Sight	Iron sight, Optional optics.

FIGURE 4.16 AN 94.

AN 94

The special features of this weapon that it can fire 1,800 rounds per minute **in a two-round burst mode,** and a special gas–recoil combo operation mechanism has been provided to achieve the feat. It may be noted that whereas in the AK-107 is a selective-fire weapon, with a **three-round burst capability** in addition to the semi-automatic and fully automatic firing modes. Even if only one or two rounds have been fired, the system on the AK-107 resets to three-round burst every time the trigger is released. The rate-of-fire is listed at 850 rounds per minute. This firing rate would have been impractical with the 7.62 × 39-mm ammunition (Table 4.7 and Figures 4.16–4.19).

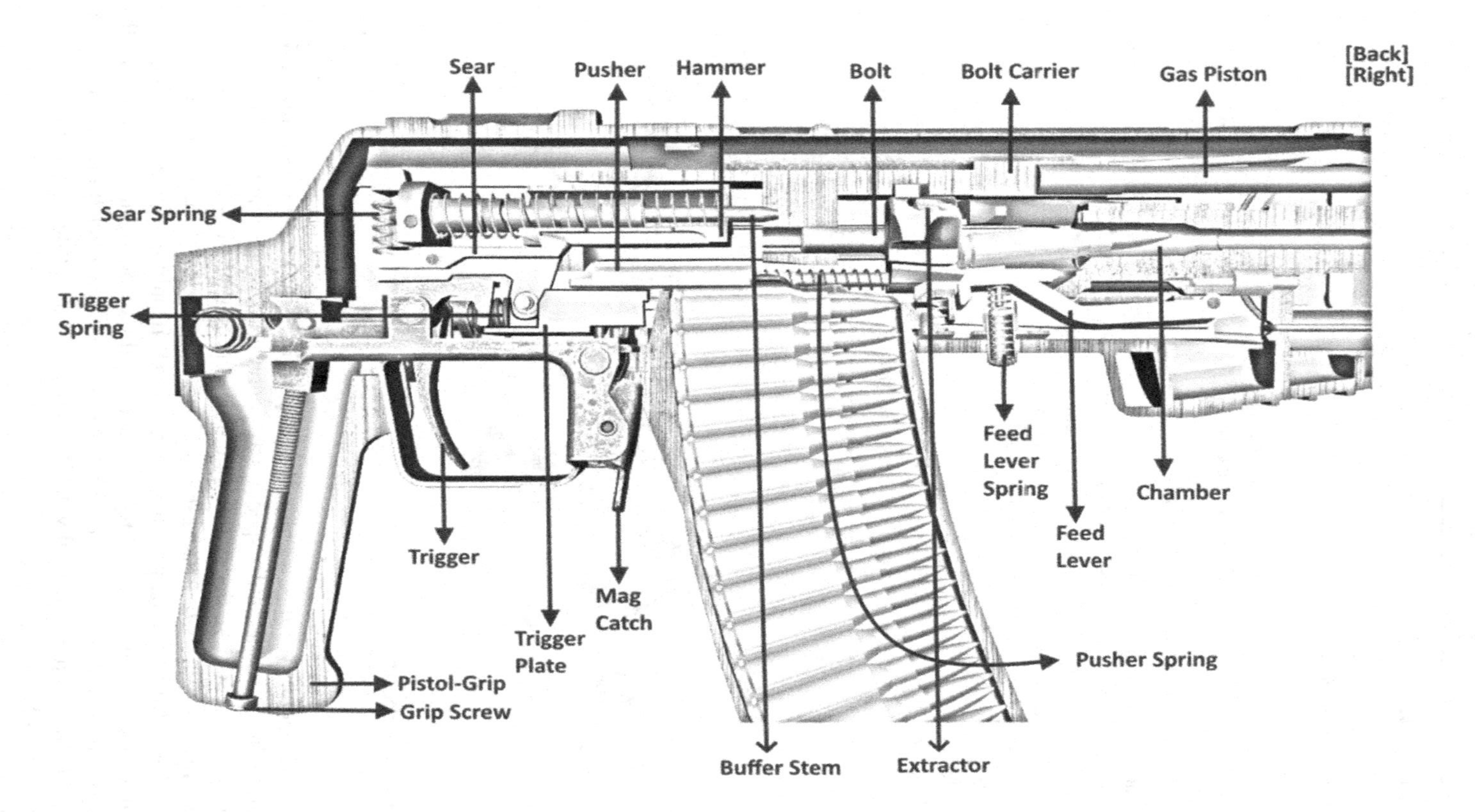

FIGURE 4.17 Right side receiver view.

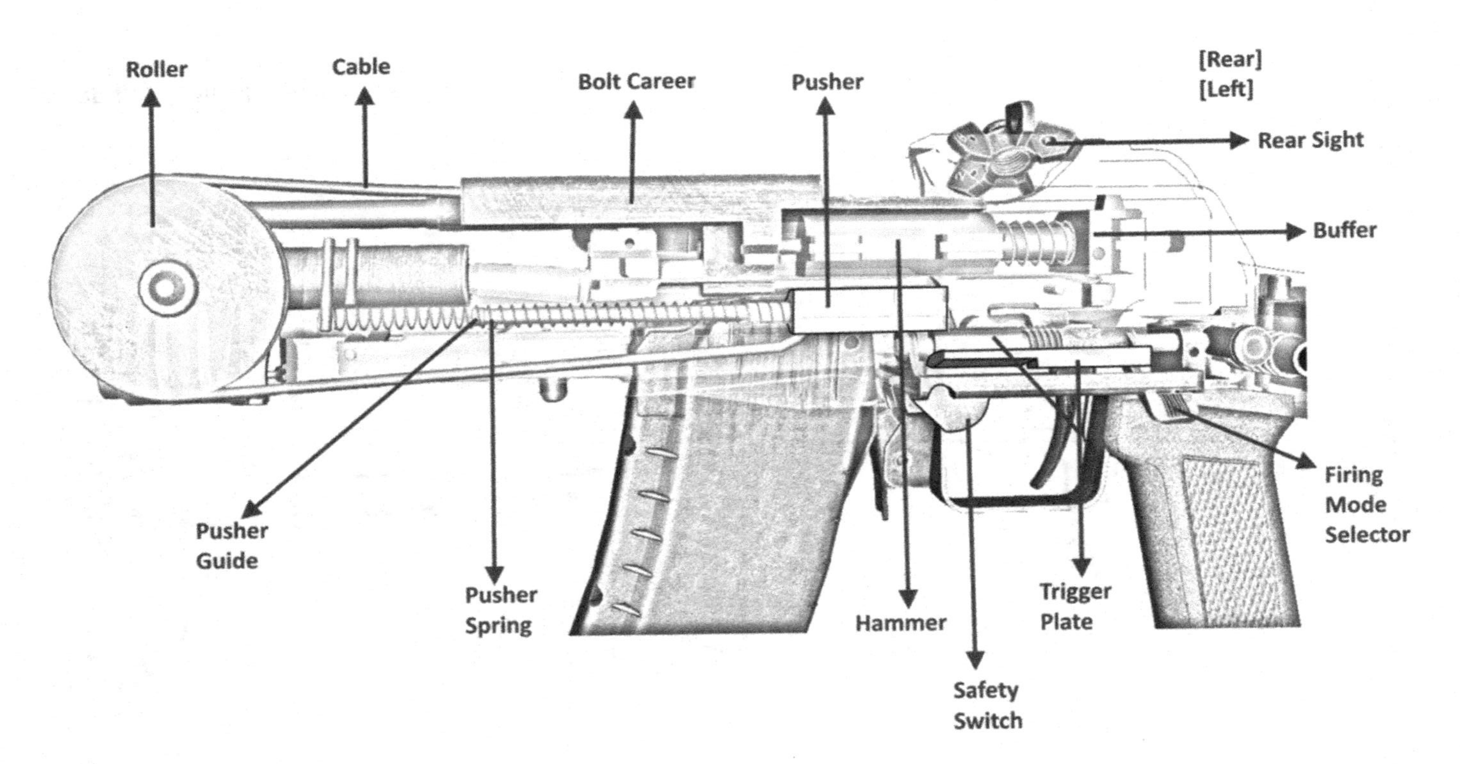

FIGURE 4.18 Left side receiver view.

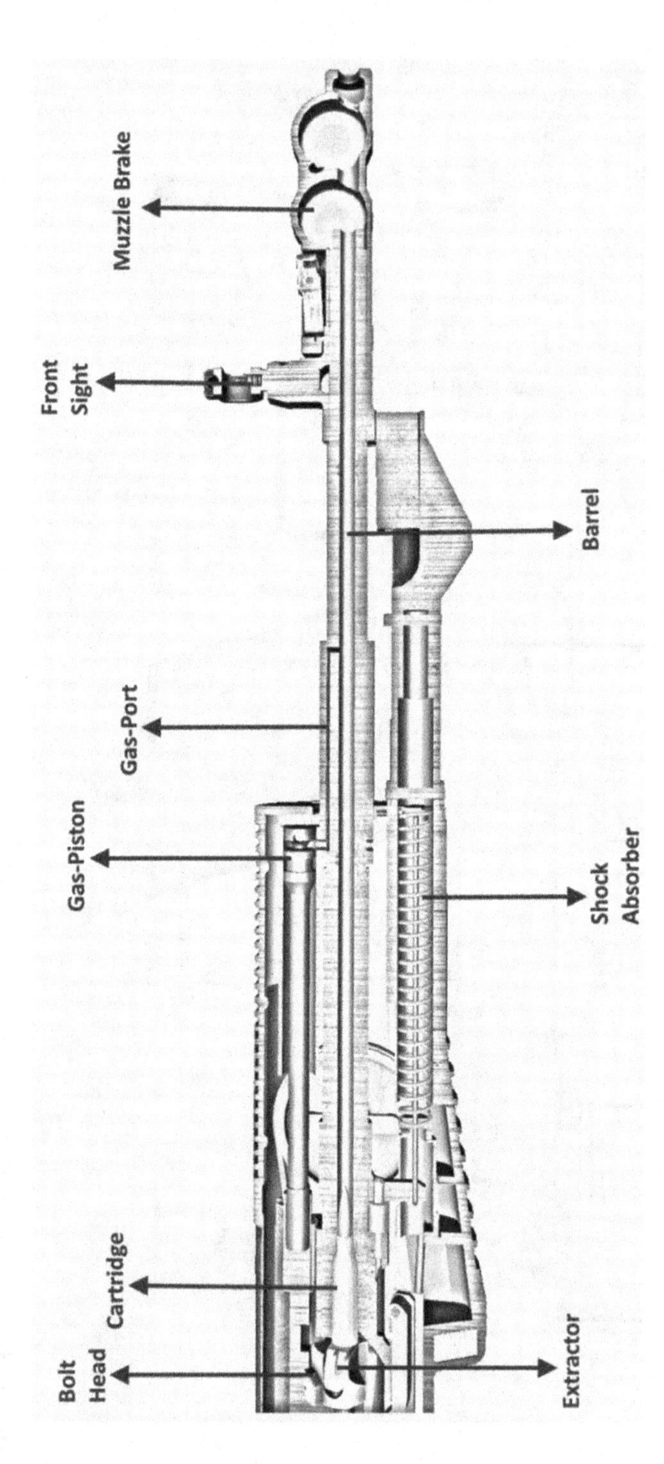

FIGURE 4.19 Left side barrel cross section.

RUGER PRECESSION RIFLE (TABLE 4.8 AND FIGURES 4.20–4.23)

TABLE 4.8
Ruger Precession Rifle**

General Specification	
Type	Marksman/Sniper rifle
Place of Origin	USA
In-Service	2015–present
Designers	Strum, Ruger & Co.
Weight	9.8 lb. for 0.308 Win version.
Length	39.25–46.75"
Barrel Length	20–26"
Twist Rate	1:7–1:10"
Cartridge	6 mm to 0.308 Win
Action	Bolt action
Rate of Fire	1,800 RPM (2 rd Burst) 600 RPM (Full Auto)
Muzzle Velocity	2500–2700 ft/s
Effective Range	4,800 ft
Feed System	10-rounds detachable box magazine
Sight	Optic scope.

FIGURE 4.20 Ruger precession rifle.

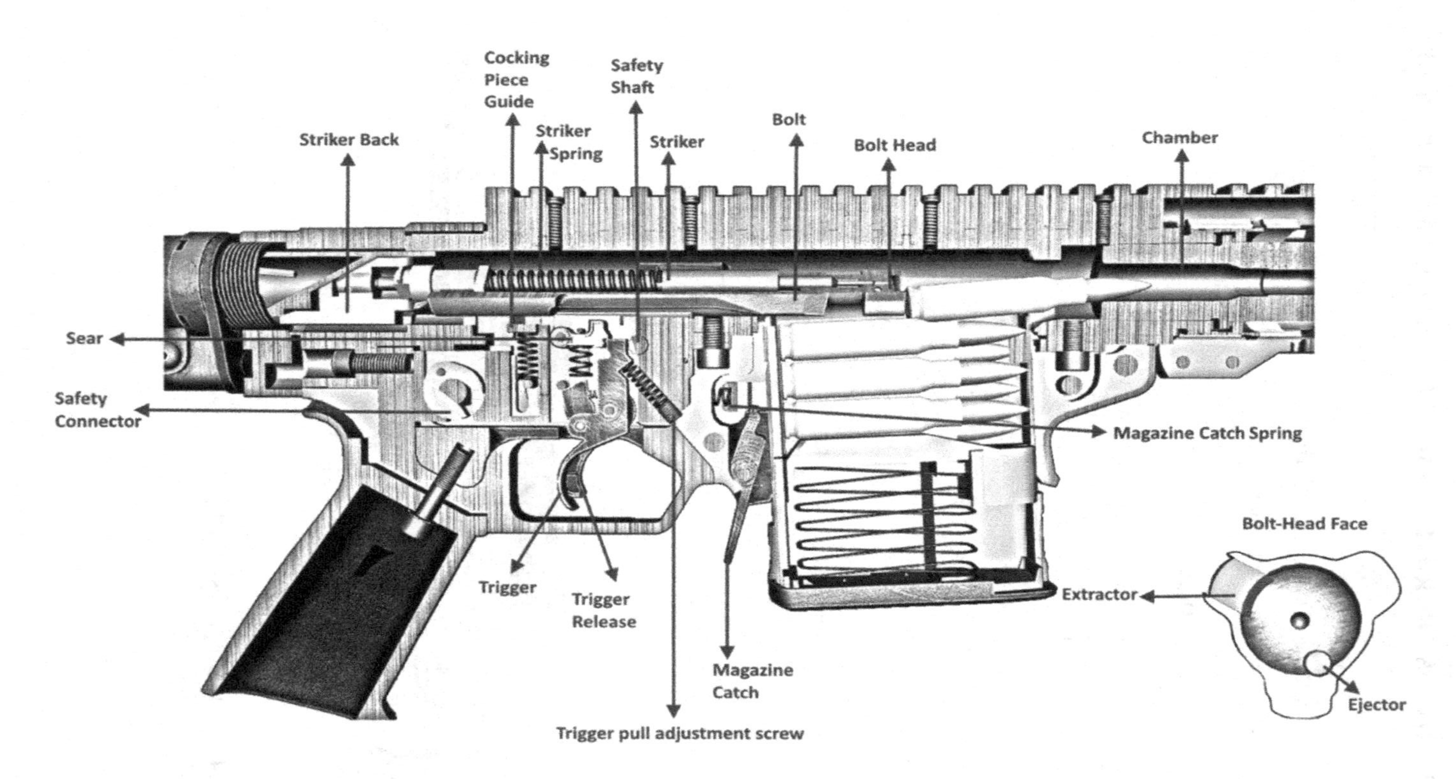

FIGURE 4.21 Right side view (trigger mechanism) lock.

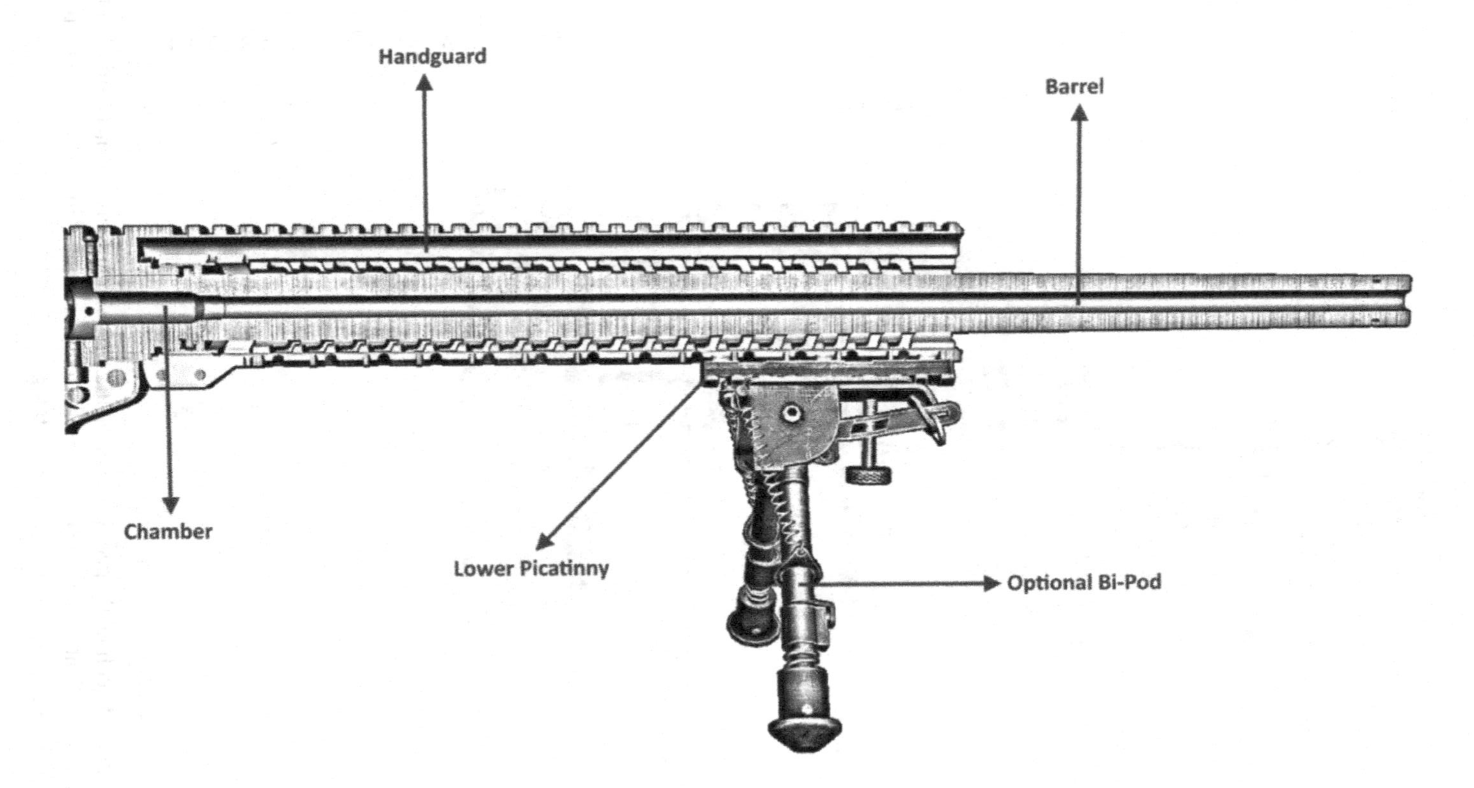

FIGURE 4.22 Right side barrel view.

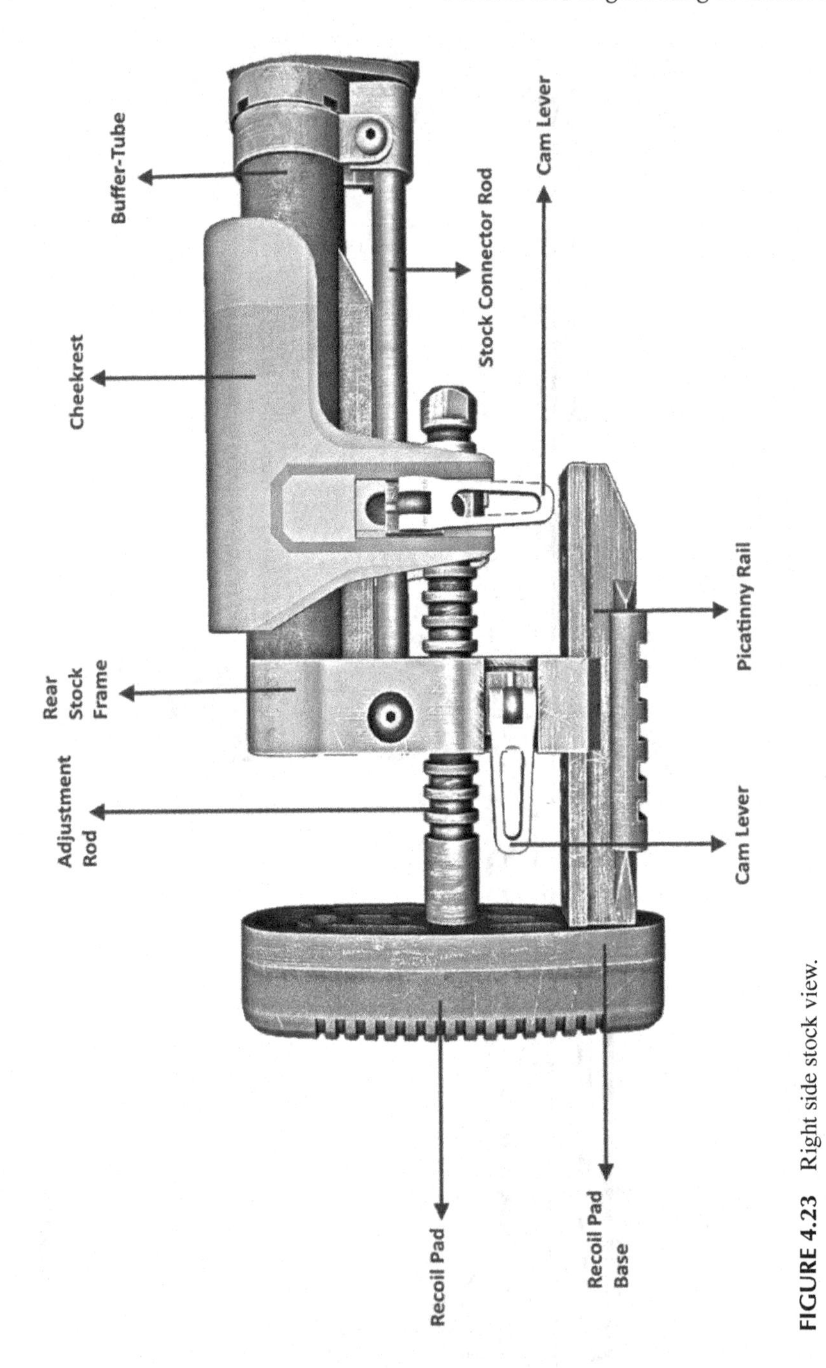

FIGURE 4.23 Right side stock view.

**The table of technical specifications of all the weapons as given in this book are tentative, indicative technical characteristics. For exact and detailed value, the readers may refer to the manufacturer's catalog.

5 Basic Design Concepts

5.1 TRIGGER MECHANISM

The trigger mechanism or action is the heart of the weapon firing mechanism. The trigger mechanism essentially consists of a trigger, sear, auxiliary sear, safety sear, a firing mode selector, and a safety lever that finally controls the spring energized striker or the hammer to regulate the firing sequence of automatic/semi-automatic firearms in general. It may sometimes include an intermediate link to transmit the initial trigger action to translate into the final action of the hammer/striker. The springs employed in the trigger mechanism also play a crucial role in the flawless functioning of the weapon. Depending on the type of weapon, the trigger mechanism may also consist of an indexing mechanism and firing rate limiter for burst and automatic control. Each component of the trigger mechanism is of a very complex shape. In terms of dimensional and inertial precession, the mechanism parts demand precise quality control and exact metallurgical properties in all its components, including the spring. Depending on the type of weapon, the number of components, and the kinematic composition of the trigger mechanism, the kinematic system varies. The manufacturing methods and the quality control of trigger mechanism components are of utmost techno-economic importance in the production of firearms parts. The options of various available manufacturing technologies are weighed carefully in taking decisions about the routes of manufacture of all trigger mechanism parts. Several possible configurations of the trigger mechanism are discussed below.

TRIGGER

- The trigger is a mechanism responsible for actuating the firing sequence of a firearm.
- When a small amount of energy is applied to the trigger, it causes much more energy to release.

Function

- In Firearms, triggers are used to initiate the firing of a cartridge seated within the gun barrel chamber.
- This is performed by activating a striking device through a system of:
  a) Spring (which stores elastic energy)
  b) Trap mechanisms that are created with a sear/trigger, sear/hammer interface that holds the spring under tension

DOI: 10.1201/9781003199397-5

c) An intermediate mechanism to transmit the kinetic energy from the spring release
d) A firing pin which strikes and ignites the primer

Types of Striking Mechanisms

- Hammer and striker are the two main types of striking mechanisms.
- The hammer is a highly complex-shaped metallic part of precise dimensions and mass. It is pivoted on a pin and is directly subjected under spring tension, and when it is released, it rotates swiftly to strike the firing pin with a sharp blow.
- The energy of the hammer delivers an impulse to the firing pin that moves along the longitudinal axis and then strikes the cartridge primer, igniting the propellant charge within the cartridge case, and the result is the subsequent discharge of the bullet.
- A striker is a direct spring-actuated firing pin. The initial stored energy of the compressed spring of the striker releases the energy for igniting the primer; it eliminates the need to be struck intermediately by a separate hammer.
- The sear surface is the trap interface between the trigger and the hammer/striker.
- There are variations in the striker mechanisms. These might have trap surfaces directly on the trigger and hammer or may have separate sears or other connecting parts.

Actions (Mechanisms)

- The firing mechanisms are also known as actions, which may be of several types of combinations of constituent parts like the trigger, hammer, and safeties that are considered as a unit or the logic of how it is built and how it is used.
- They are categorized according to which functions the trigger is to perform.
- The trigger releases the hammer or the striker, as well as it may cock the hammer or striker, deactivate passive safeties, rotate a revolver's cylinder, select between semi-automatic, burst, and full-automatic fire such as the Steyr AUG (progressive trigger), or pre-set a "set trigger".

A: Single-action

- The earliest and mechanically simplest type of trigger is the single-action (SA) trigger.
- It is called the "single-action" because it performs a *single action* that is releasing the hammer or striker to discharge the firearm every time the trigger is pulled.
- In firearms having the SA trigger mechanism, the hammer must be cocked separately.

- In modern days, the terms "single-action" and "double-action" almost always refer to handguns, as very few if any rifles or shotguns have double-action triggers.

B: Double-action Only

- A double-action, also known as double-action-only (DAO), is a design that has no internal sear mechanism to hold the hammer or striker in the cocked position (semi-automatics). It has the entire hammer shrouded. It has the thumb spur machined off to prevent the user from cocking the revolver.
- For every single shot, the DAO design requires a higher trigger pull to both cock and trip the hammer.
- The DAO action in a semi-automatic is to eliminate the change in trigger pull between the first and subsequent shots that one experiences in a DA/SA pistol while avoiding the perceived danger of carrying a cocked handgun which is single-action.

C: Double-action/Single-action

- A double-action/single-action (DA/SA) firearm has a combination of the features of both mechanisms.
- When we refer to DA/SA, it generally implies a semi-automatic firearm.
- In a revolver, "double action" means a handgun having the ability to fire both in double- and single-action mode. A revolver can also have plain single-action only.
- "Double-Action" refers to a gun trigger mechanism that cocks the hammer and then releases the sear, thus performs two actions, so it is *double action*. Most "double-action" guns have both single- and double-action abilities.
- "DAO", or "Double Action Only" firearms, cannot fire in single-action mode.
- In a DA/SA semi-automatic pistol, the trigger mechanism functions in the same way as that of a DA revolver. Most semi-automatics can self-cock the hammer when firing.
- It is possible to carry the weapon with a loaded chamber and hammer down, thus the danger of carrying a single-action semi-automatic weapon in loaded condition can be reduced. And when required the user can fire the weapon in double-action mode simply by pulling the trigger.
- Because of double-action mode, the semi-automatics cycle the slide to automatically cock the hammer to the rear, so after the first shot the rest of the shots will be fired in single-action mode.
- This is the positive aspect of a single-action trigger where there is no need to carry a weapon cocked and locked (with a loaded chamber and cocked hammer), or with an unloaded chamber.

TYPES OF TRIGGER

A: Release Trigger

When a hammer or striker is released by releasing the trigger instead of pulling it such type of trigger is called a release trigger. Remington 870 shotgun incorporates such a trigger (Figure 5.1).

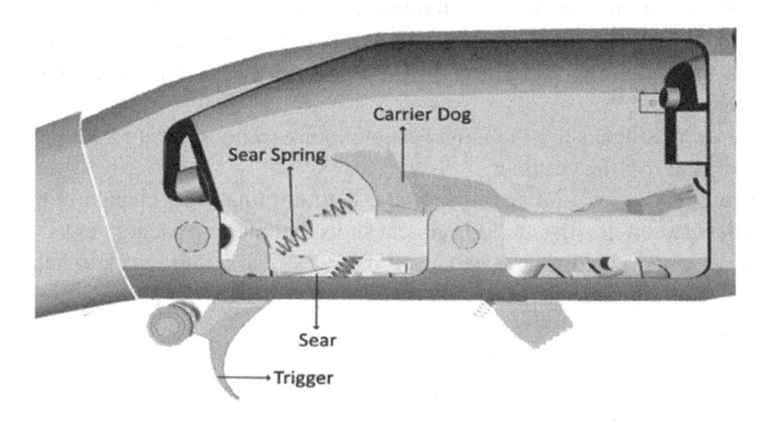

FIGURE 5.1 Remington 870 shotgun.

B: Binary Trigger ("Pull and Release")

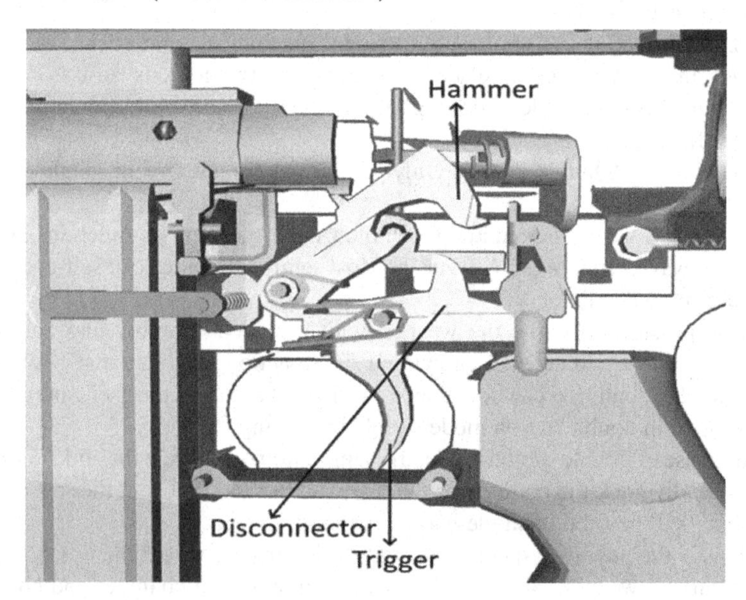

FIGURE 5.2 LR300.

A binary trigger is a trigger for a semi-automatic firearm that drops the hammer both when the trigger is pulled and when it is released. Examples include the AR-15 series of rifles. The AR-15/LR300 trigger, as produced by Liberty Gun Works, only functions in pull and release mode and does not have a way to catch the hammer on release (such triggers are not permissible by arms law) (Figure 5.2).

Set Trigger

Example of Set Trigger: Remington 700

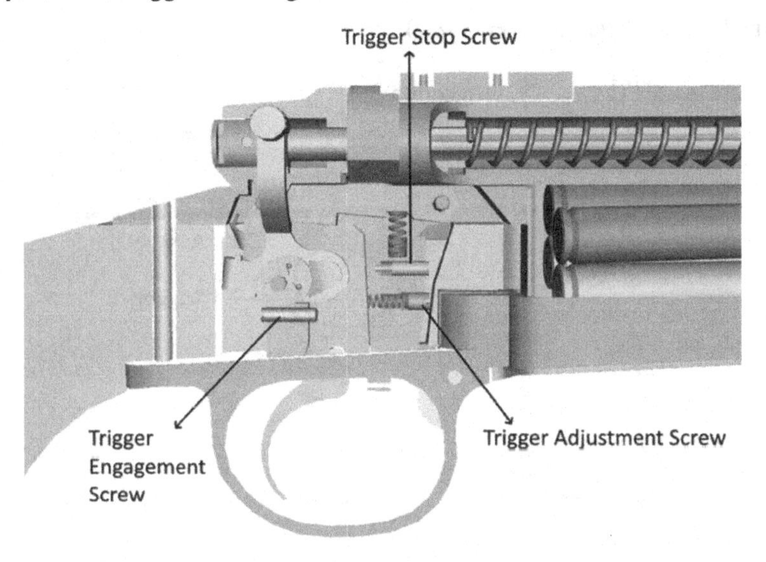

FIGURE 5.3 Remington 700.

With a set trigger, a shooter can adjust the trigger pull (the resistance of the trigger) to suit his need and maintain the required degree of safety in the field in comparison to a conventional, very light trigger. These are available in two types: single set and double set. Set triggers are used on customized weapons and competition rifles where the shooter feels a light trigger pull is beneficial for accuracy (Figure 5.3).

A: Single Set Trigger

- A single set trigger usually requires a conventional amount of trigger pull weight or it is "set" – by directly pushing forward on the trigger, or by pushing a small lever attached to the rear of the trigger.
- Because of allowing slack in the trigger, a much lighter trigger pull is felt. This is normally known as a hair-trigger.

B: Double Set Trigger

- A double set trigger uses two triggers: one sets the trigger and the other shoots the weapon.

- Double set triggers are classified by phases.
- A double set, single-phase trigger, which can only be operated first by pulling the set trigger and subsequently pulling the firing trigger.
- If the set trigger is not pulled a double set, double phase trigger can be operated as a standard trigger, otherwise as a set trigger by pulling the set trigger first.
- A double set, double phase trigger has the advantage of both a set trigger and a standard trigger.

Pre-set Trigger (Striker or Hammer)

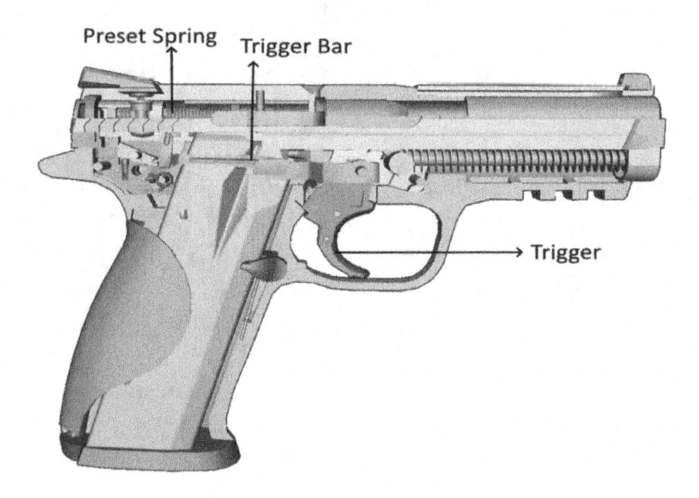

FIGURE 5.4 S&W M&P .40 pre-set striker.

- Semi-automatic handguns only use pre-set strikers and hammers.
- The hammer or striker rests in a partially cocked position after firing a cartridge or loading the chamber.
- The trigger performs the function of initiating the cocking cycle and then it releases the striker or hammer. While technically it amounts to two actions, it is different from a double-action trigger in that the trigger cannot fully cock the striker or hammer.
- It is different from single-action in the aspect that if the striker or hammer were to release, it would generally not have enough power to detonate the primer.
- Glock, Smith & Wesson M&P, Springfield Armory XDS (only) are examples of pre-set strikers. These are also called Striker Fired Action or SFA (Figure 5.4).

Pre-set Hybrid Trigger

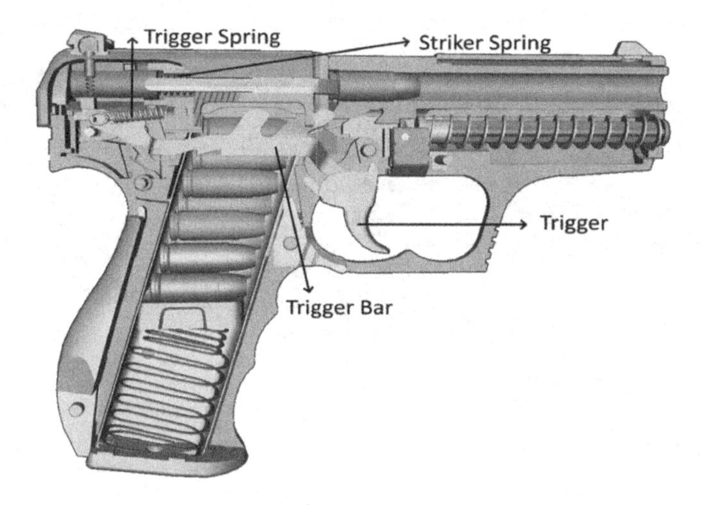

FIGURE 5.5 Walther P99 AS.

- If we reverse the DA/SA trigger the result is the pre-set hybrid trigger. The trigger is pre-set by the first pull and if the hammer or striker fails to detonate the cartridge, the trigger can be pulled again and thus operate as a double-action-only till either cartridge discharges or the malfunction is rectified.
- Thus, the operator has an opportunity to attempt a second time after a misfire/malfunction while in single-action if a misfire occurs the only option is to rack the slide to clear the round and re-cock the hammer.
- Thus, it is advantageous in a sense that many rounds may fire on striking the second time and it is also faster to trigger the second time than racking the slide to clear the misfire round and fire a fresh round.
- But it is disadvantageous if the round fails to fire on the second strike, because the user will be forced to clear the misfired round, thus wasting time than if he had simply done so in the first place (Figure 5.5).

Variable Triggers

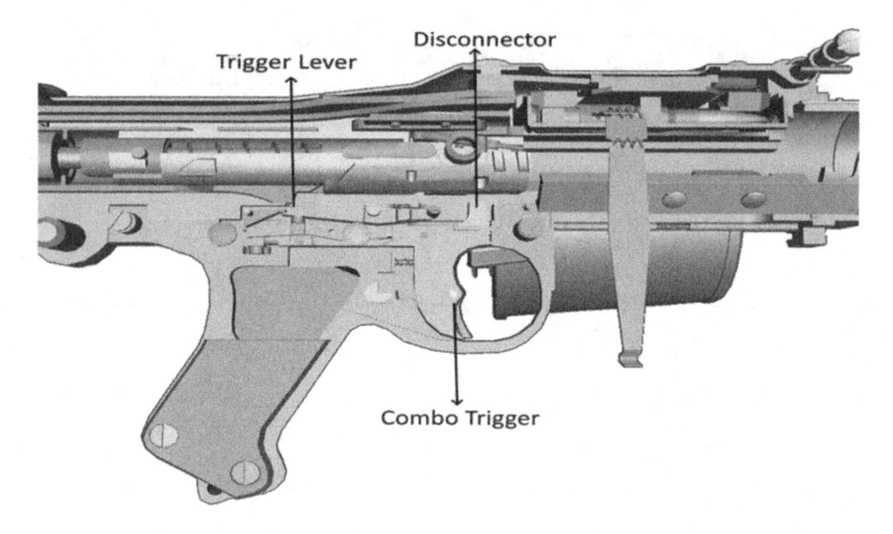

FIGURE 5.6 MG 34.

A: Double-crescent Trigger (Variable)

- There is no need for a fire selector in a double-crescent trigger. To produce the semi-automatic fire, the upper segment is pressed and the fully automatic fire is produced by pressing the lower segment. The idea is innovative but is not normally used because of its complexity. However, MG34 and M1946 automatic rifles feature such a device (Figure 5.6).

Progressive/Staged Trigger(Variable)

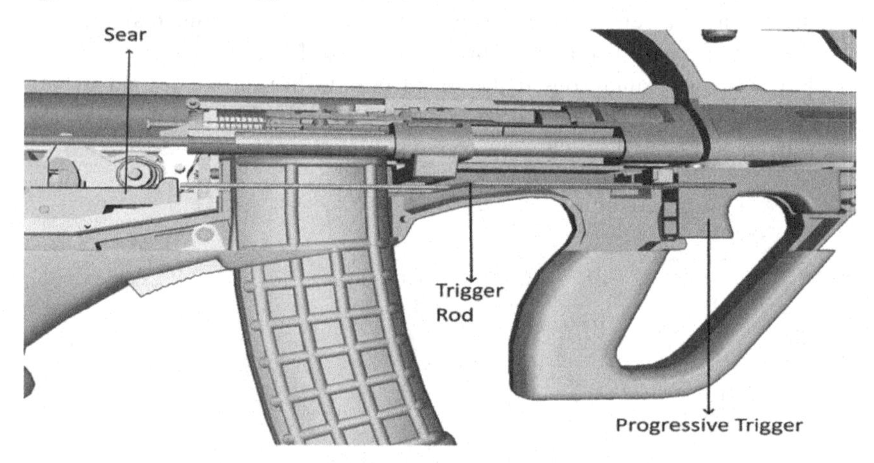

FIGURE 5.7 Steyr AUG v2 – Progressive/staged trigger.

- There is no fire selector in a progressive or stage trigger. The mechanism is so designed that the extent of depression of the trigger gives a proportional firing rate. When the trigger is pulled lightly, the firearm produces a single shot and when the trigger is depressed further, the weapon produces a fully automatic firing. FN P90, CZ Model 25, PM-63, and Steyr AUG have such a progressive trigger mechanism (Figure 5.7).

Trigger Pull Stages

The trigger pull can be divided into three distinct stages:

a) Take-up or pre-travel: This is the movement of the trigger which occurs before the sear starts moving.
b) Break: The movement of the trigger after which the sear reaches the point of release.
c) Overtravel: The extra movement of the trigger after the release of sear.

A: Take-up

- Take-up allows a distinct movement of lighter pull before the break. The weapon having this feature in a noticeable amount is considered to have two stages, because a distinct increase in force is required to pull the trigger to cause a break after the take-up.
- When there is no noticeable movement before the break, the trigger is said to be a single-stage trigger.
- The trigger may be designed in a fully adjustable way to function either as a single-stage or double stage by manipulating the take-up.
- Two-stage triggers are often called pressure triggers whereas, a single-stage trigger is also known as a direct trigger.

B: Break

- The break is the moment just before the shot being struck by the striker. The break may be soft and smooth with some amount of trigger travel in the firing. Some designs provide for an abrupt and forceful break with little trigger movement.

C: Overtravel

- The overtravel is the extra travel of the trigger after the shot is fired and it is important when there is a sudden release of large resistance as the sear breaks. Normally, an overtravel stop is provided to control the motion of the trigger just after the break.

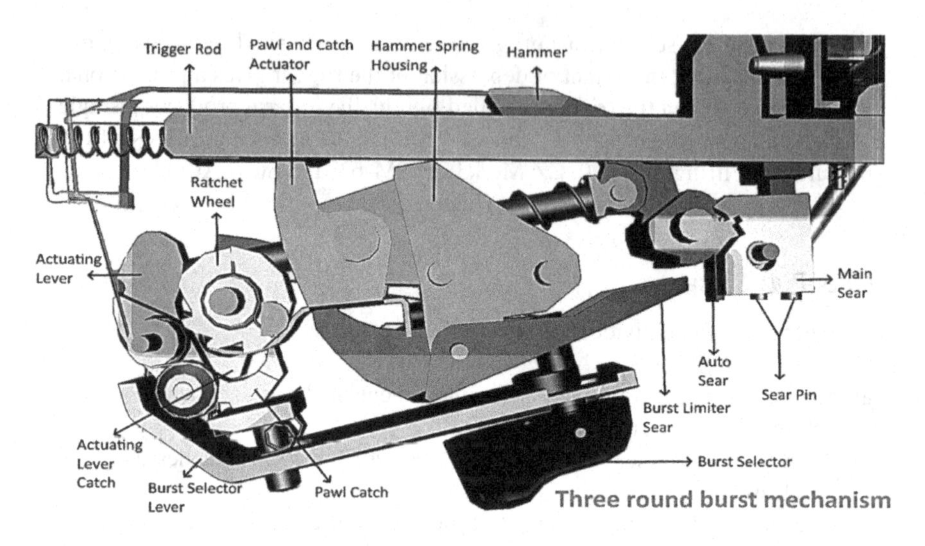

FIGURE 5.8 Three-round burst mechanism.

Burst Mechanism

It is used to control the number of rounds fired in a single trigger pull. Normally it is available as a two-round or three-round burst. The above figure is a schematic of a three-round burst mechanism (Figures 5.8 and 5.9).

Essential Parts:

- **A Burst selector** which positions the burst selector lever
- **A burst limiter sear** which only comes into action in the burst mode
- **A ratchet wheel and pawl catch** to count the number of rounds and automatically reset the burst mode

Note the three pairs of sear surfaces with the hammer for action in single auto and burst mode. The burst limiter sear comes into action only in burst mode.

This mechanism is incorporated in the assault rifle for controlling ammunition consumptions in auto mode as well as an increase in the hit probability in assault mode.

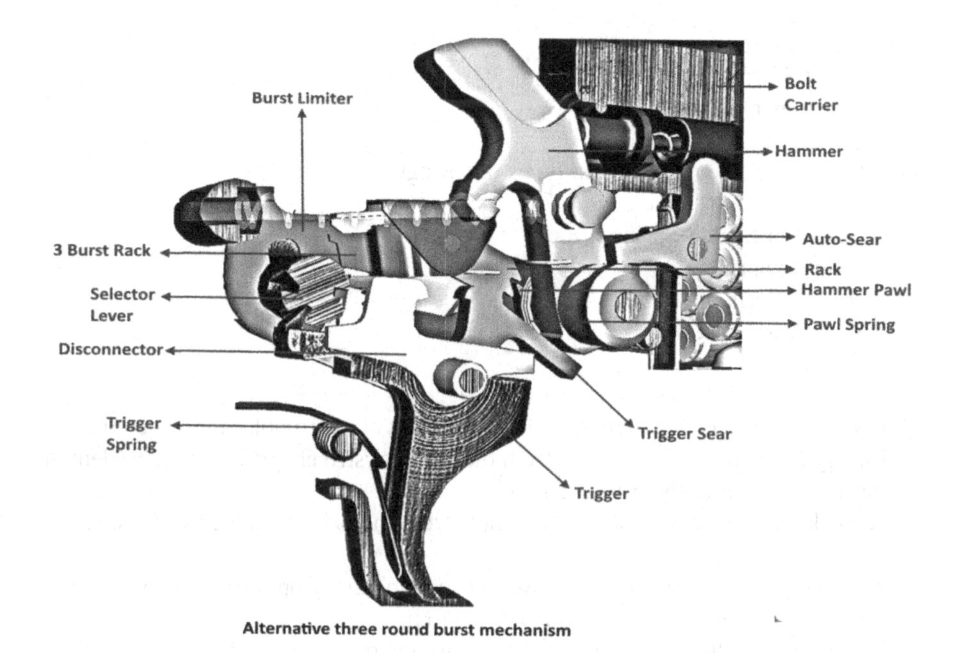

FIGURE 5.9 Three-round burst mechanism.

5.2 FIREARM SPRINGS

The springs in the firearms are critical for successful and flawless functioning. A firearm is as good as the quality of springs with which it is furnished. The springs are used in several configurations as helical compression spring, torsion spring, leaf spring, buffer spring, detent, wire-forms, etc.

Components of which springs are used as follows:

- Magazine Springs and Magazine Catch Springs
- Recoil Springs
- Trigger Springs
- Hammer Springs (mainsprings)
- Action Returns
- Anti-Rattle Springs
- Bolt Springs
- Ejector Springs
- Extractor Springs
- Firing-pin/Striker Springs
- Plunger Springs
- Safety Pins
- Sear Springs

- Selector Springs
- Sight Springs
- Takedown Springs

Of all the above, the most critical types of springs:

1. Hammer/Striker Spring
2. Trigger Spring
3. Recoil Spring
4. Extractor Spring
5. Magazine Spring

If they're damaged or lose strength, they give most of the troubles in the action.

Energy to fire the primer comes from the hammer/striker springs. A sharp dent in the primer cup ignites the primer charge.

A weak hammer/striker spring may not strike the primer with enough power to ignite it.

The strength of the spring may have to overcome many opposing forces to ignite the primer. So, it should have adequate force to fire the round.

The purpose of the trigger spring is to return the trigger to the pre-firing position. If the return trigger spring fails, the trigger doesn't reset the weapon, and it becomes nearly inoperable.

Recoil spring is the source of energy for self-loading and automatic firearms. This spring slows down the recoil assembly of the bolt and bolt carrier and returns the spring into the battery after firing. This is a helical compression spring and the largest of the springs in the weapon. If the spring is too weak, the recoiling carrier and breech assembly will batter the receiver of the firearm or it may even fall short of power to bring the weapon fully back into the battery. The friction in the fully loaded magazine also demands enough force in the spring to get into the battery. Conversely, if the recoil spring is too strong it may cause the weapon to short stoke and cause the problem of stovepipe and other jams.

There are several geometries and designs of the extractor, but they all have the purpose of activating the extractor, so that the spent shell casings are removed from the chamber and thrown out of the weapon with the help of an ejector. If the spring is too weak, the weapon exhibits failures to extract or eject with an improper ejection pattern. If this spring is too strong, the weapon will give trouble in going into the battery.

Magazine Springs perform the function of placing the next cartridge into position as soon as the bolt-breech assembly crosses the magazine in the recoil stroke. Since the recoil process if very fast, the next round must be placed into position immediately after the breech assembly has crossed the magazine, and if it doesn't happen, the block breech and carrier will be in the battery, without taking out any cartridge. A weak or worn magazine spring may cause multiple malfunctions. If a magazine spring is very hard, it can cause damage to the feed-lip, or slow the dynamic parts like carrier, block breech, and the slide, by excessive friction so much so that they may fail to function properly.

Springs made from music wire start degrading the moment it is assembled and continues to degrade with each action. A striker spring has been found to show up to a 30% loss in power within one year after assembly, even without being used. There is just no way a gun spring in this condition can produce the energy required for proper and consistent function.

A silicon–chrome gun spring which is made from premium wire and pre-set gives hundreds of thousands of consistent cycles and does it with less vibration and sight disruption.

All gun springs are recommended to be made from silicon–chrome or high-grade alloys wire that has been cold drawn and heat treated. No guns are better than the quality of the springs with which they have been assembled (Figure 5.10).

5.3 BARREL

Design of gun barrel broad objectives:

1) **Design for performance to ensure:**
    - Optimum muzzle velocity of the projectile
    - Improved the surface-to-volume ratio for heat dissipation
    - To make the barrel more efficient to cool after firing
    - Efficient burnout of propellant
    - Minimum perceived recoil
    - Maximum transfer of energy of bullet to the target
    - Maximum possible accuracy; thus, maintain stability in flight
2) **Check Design for strength**

    Safety in gun design is of paramount importance and the strength should be adequate to:

    - To prevent catastrophic failure and explosive fracture
    - To retain the structural strength, rigidity, and geometry
    - To increase the overall specific strength;
    - To ensure this, various international standards specify the appropriate material and minimum wall thickness which are mandatory

Design for Strength

Any parts of the weapon fail on account of the crossing of the limits of the following factors which the gun material is capable of enduring:

Failing factors are Stress, Strain, and Energy that the material is capable of withstanding. Academic theories which are available for analysis of the failure are as under:

- Maximum principal stress theory (Rankine)
- Maximum principal strain theory (Tresca)
- Maximum strain energy density theory (Haigh)

- Maximum distortion energy density theory (Von mises)
- Principal strain theory
- Maximum strain energy theory and shear strain energy

The various failure theories are dependent on the nature of the material for its applicability of obtaining the best inference and optimum dimensional design. E.g., the designs of parts made of brittle materials are tested by application of any of the following theories:

- Maximum principal stress/strain
- Maximum strain energy

Similarly, for ductile materials, the testing theories for design adequacy are tested by:

- Maximum shear stress theory
- Maximum distortion energy density

In the case of firearm parts, most of the parts are so designed that they work with stress almost near to their maximum bearable stress. The weapon parts are also designed for just enough materials required for satisfying the mechanical impedance and strength. The components also are precise parts to meet kinematic relationships for flawless operation. Even slight distortion or wear in some critical parts leads to unacceptable failure of the weapons. In general, the test results are available for materials either in simple tension or in compression. The various failure criteria are examined by comparing their condition under the complex stress condition with the failure criteria in a simple tension test and a factor of safety is thus determined (Figure 5.10).

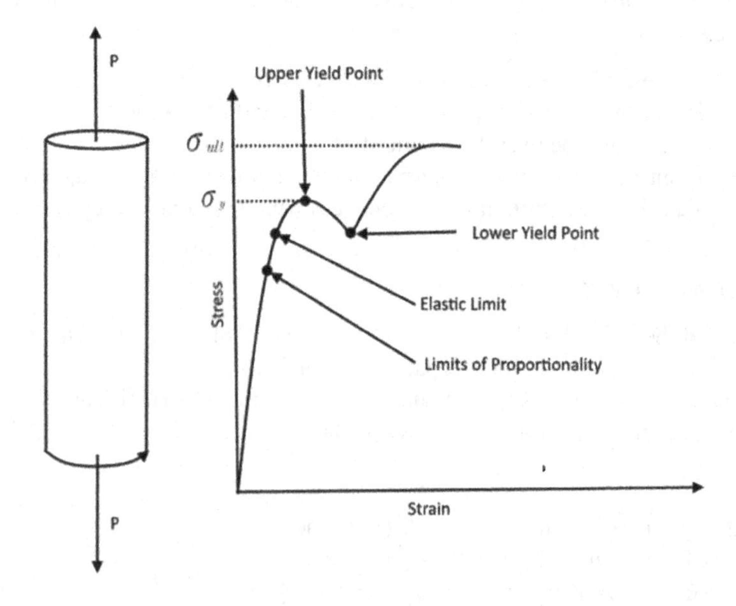

FIGURE 5.10 Simple tensile test character of ductile material.

Failure is considered to have taken place when:

$$\sigma_1 \geq \sigma_{ult} \text{ (Rankine)} \qquad \text{Eq 5.1}$$

$$\text{Strain}_{max} = \frac{\sigma_y}{E} \text{ (Tresca)} \qquad \text{Eq 5.2}$$

$$\text{Strain Energy Density} = \frac{\sigma_y^2}{2E} \text{ (Haigh)} \qquad \text{Eq 5.3}$$

$$\text{Maximum distorsion energy density} = \frac{\sigma_y^2}{6G} \text{ (Von Mises)} \qquad \text{Eq 5.4}$$

where σ_y = yield point strength, E = Modulus of elasticity

G = Modulus of rigidity of the material, σ_1 = Principal stress, σ_{ult} = Ultimate strength

A comparison of design strength is given below to appreciate the sizes and safety margins that may be available from designs by various theories. It should be amply clear that design by application of Von Mises results in the optimum design.

Design by the application of Haigh theory (Red) will result in the most conservative design. The calculated size that may be expected by the application of Tresca (Green) and the Haigh (Red) will result in nearly of comparable size (Figure 5.11).

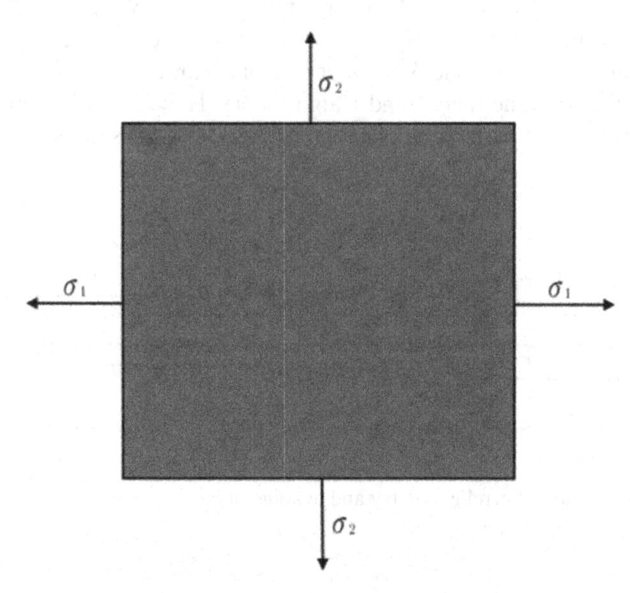

FIGURE 5.11 An element subjected to two-dimensional principal stresses.

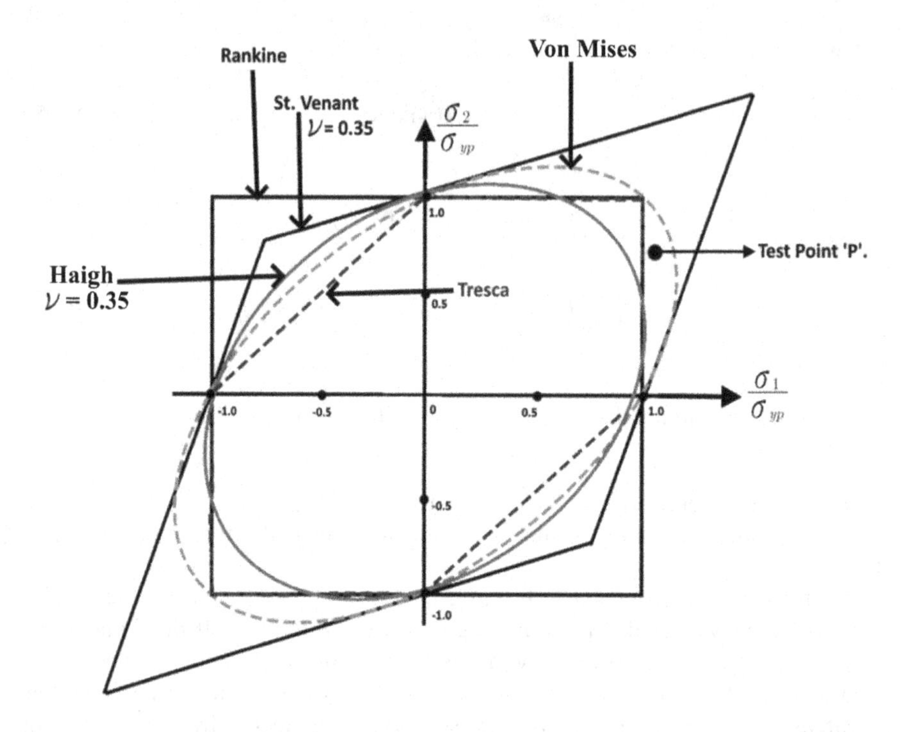

FIGURE 5.12 Comparison of failure theories.

Interpretation of test point 'P' – An element subjected to two-dimensional stresses is considered to be within the safe limit of design when considered by the application of St Venant's theory and Von Mises theory. However, it will be considered unsafe from the Rankine theory and Haigh theory. However, it is well established that design of firearm parts by application of Von Mises is safe and cost-effective (Figure 5.12).

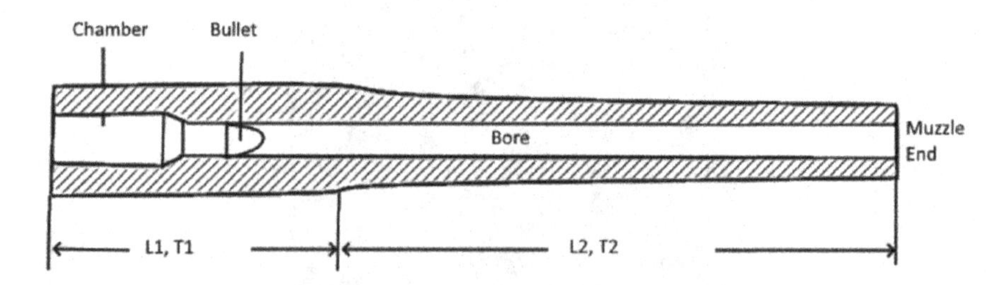

FIGURE 5.13 General barrel geometry and average pressure.

$$P_{av}(T1) = \frac{1}{T_1}\int_0^{T_1} p(t)\,dt, \text{ Section 1} \qquad \text{Eq 5.5}$$

$$P_{av}(T2) = \frac{1}{T_2}\int_0^{T_2} p(t)\,dt, \text{ Section 2} \qquad \text{Eq 5.6}$$

Design Pressure Calculation Dynamic Factor (Kd) See Figure 5.13

Rate of Fire	Section L1	Section L2
< 30 R.P.M	3.0	2.5
500-600 R.P.M	3.5	3.0

- Design pressure= Dynamic factor (Kd) X Average pressure (Figures 5.14 and 5.15).

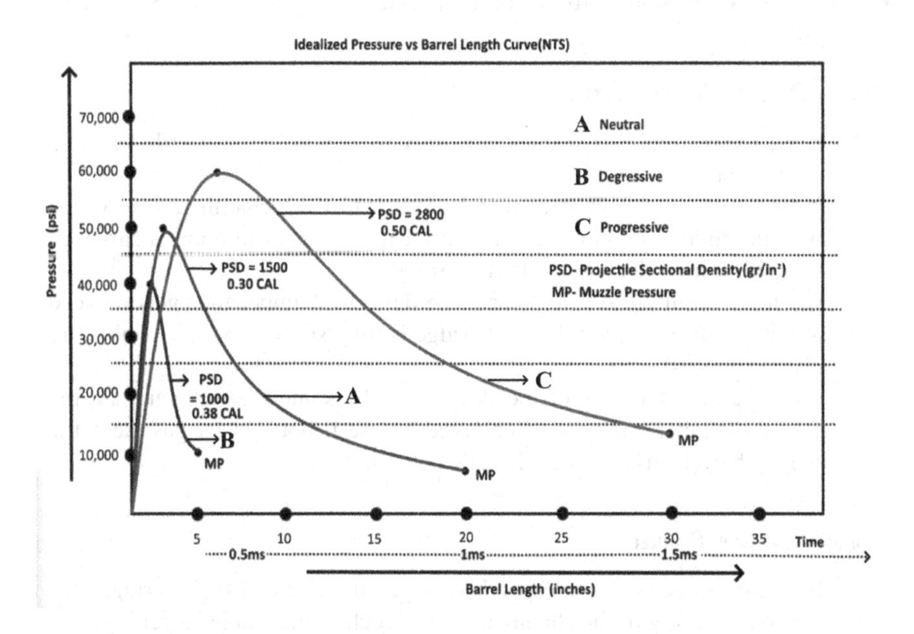

FIGURE 5.14 Idealize pressure vs barrel length curve NTS (Not to scale).

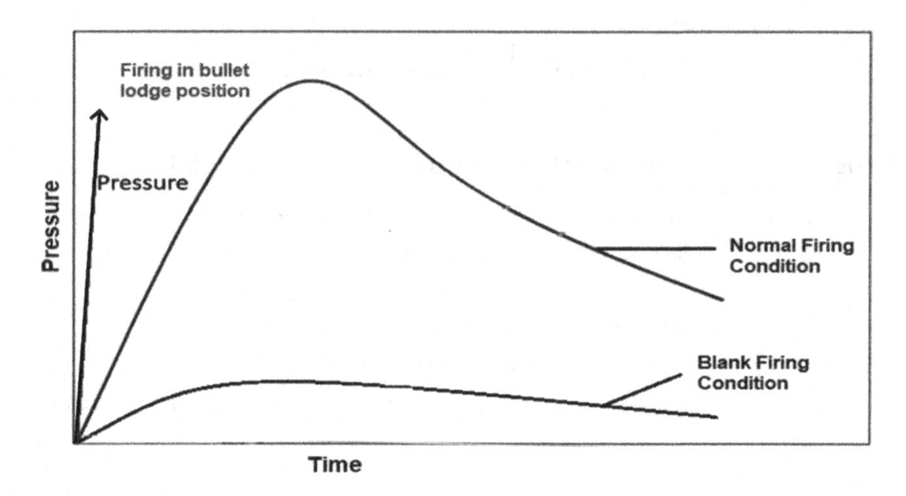

FIGURE 5.15 General pressure vs time graph. General nature of pressure–time curve in a gun barrel.

Barrel Design Consideration

- The chamber and crown must smoothly transition the projectile into and out of the rifling.
- Rifling may not begin immediately forward of the chamber. The design may incorporate an unrifled throat after the chamber to home a cartridge without pushing the bullet into the rifling.
- The force required to load a cartridge into the chamber thus gets reduced and the chances of an unfired cartridge getting stuck into the chamber are also minimized.
- The specified diameter of the throat may be somewhat greater than the groove diameter and may be enlarged by use if hot powder gas melts the interior barrel surface when the rifle is fired.

Components of Barrel

- The chamber contour corresponds to the casing shape of the cartridge.
- The rear opening of the chamber is the breech of the whole barrel.
- The bore is the hollow internal lumen of the barrel and takes up a vast majority portion of the barrel length.
- The muzzle is the front end of the barrel from which the projectile will exit. Precise machining of the muzzle is crucial for accuracy.

Chamber Features

When a firearm cartridge is chambered, its casing, consisting of

- The body
- Shoulder
- Neck

occupies the chamber.

Bore Features

- The projectile's status of motion/internal ballistics while traveling down the barrel is determined by the bore.
- A rifled bore imparts spin to the projectile about its longitudinal axis, which gyroscopically stabilizes the projectile's flight.

Bore Details

- When a firearm cartridge is chambered, its casing occupies the chamber, but its bullet protrudes beyond the chamber into and touches the wall of the posterior end of the bore.
- In a rifled bore, this short rear section is without rifling and is called a free bore.
- The length of the free bore not in contact with the bullet is called the lead, which allows the bullet an initial "run-up" to build up momentum before encountering rifling.
- The throat is the transition zone where smoothbore changes into the rifling; this function gives the bullet a smooth lead to extrude into the rifling.
- The throat is the area which is subjected to the peak pressure and temperature and hence erodes fast. Therefore, it is the weakest area to determine barrel life (Figure 5.16).

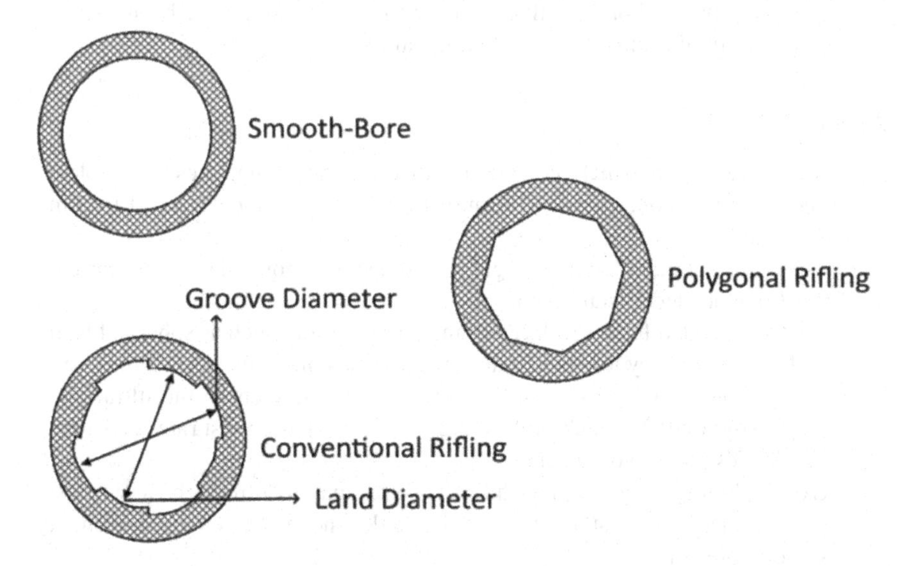

FIGURE 5.16 Barrel bore.

Rifling Design Consideration

- The common goal of rifling is to deliver the projectile accurately to the target.
- In addition to imparting the spin to the bullet, the barrel must hold the projectile securely and concentrically as it travels down the barrel.

This requires that the rifling meet several tasks

- It must be sized so that the projectile will swage or obturate upon firing to fill the bore.
- The diameter should be consistent, and must not increase toward the muzzle.
- The rifling should be consistent down the length of the bore, without changes in cross-section, such as variations in groove width or spacing.
- It should be smooth, with no scratches lying perpendicular to the bore, so it does not abrade material from the projectile.

Free Bore

- Free bore is a groove-diameter length of the smoothbore barrel without lands forward of the throat.
- Free bore allows the bullet to transition from static friction to sliding friction and gain linear momentum before encountering the resistance of increasing rotational momentum.
- The free bore may allow more effective use of propellants by reducing the initial pressure peak during the minimum volume phase of internal ballistics before the bullet starts moving down the barrel.
- Barrels with free bore length exceeding the rifled length have been known by a variety of trade names, including paradox.

Rifling

- Rifling/twist rate, which indicates the distance the rifling takes to complete one full revolution, such as "1 turn in 10 inches" (1:10 inches), or "1 turn in 254 mm".
- The combination of length, weight, and shape of a projectile determines the twist rate needed to stabilize it.
- Barrels intended for short, large-diameter projectiles such as spherical lead balls require a very low twist rate, such as 1 turn in 48 inches (122 cm).
- Barrels intended for long, small-diameter bullets, such as the ultra-low-drag, 80-grain 0.223-inch bullets (5.2 g, 5.56 mm), use twist rates of 1 turn in 8 inches (20 cm) or faster.
- Extremely long projectiles such as flechettes may require high twist rates; these projectiles must be inherently stable and are often fired from a smoothbore barrel.

Fitting the Projectile to the Bore

- In breech-loading firearms, the task of seating the projectile into the rifling is handled by the *throat* of the chamber.
- Next is the free bore, which is the portion of the throat down which the projectile travels before the rifling starts.
- The last section of the throat is the *throat angle*, where the throat transitions into the rifled barrel.
- The throat is usually sized slightly larger than the projectile, so the loaded cartridge can be inserted and removed easily.
- But the throat should be as close as practical to the groove diameter of the barrel.
- Upon firing, the projectile expands under the pressure from the chamber and obturates to fit the throat.
- The bullet then travels down the throat and engages the rifling, where it is engraved, and begins to spin.

Rifling Processes

Different rifling processes are stated as under:

- Cutting one groove at a time with a tool (cut rifling or single-point cut rifling). The starting barrel blank is a honed and polished bore finished to the land-to-land close diameter. The cutting depth of rifling is generated through the several passes of single-point cutters.
- The grooves can be cut in one pass with a special progressive broaching tool is known as broached rifling. The starting blank has a finished hole same as the land to the land diameter of the finished barrel.
- A button-shaped tool with rifled helixes can be forcefully pushed through a honed and polished bore finish to land-to-land dimension and in the process, the inside material squeezed out, creating the spiral groove. The process is known as button rifling. This is most suitable for the mass production of the pistol barrel.
- The barrel blank finished to a predetermined oversize bore is used in the hammer forging process. A set of hammers simultaneously impacts around the periphery of the blank to surge in the material over the rifled mandrel to create a reverse image of the rifling of the mandrel. The process is also known as the cold swaging process. Sometimes this is also referred to as the flow-forming process. The process is suitable for the mass production of assault rifle barrels.
- ECM and EDM are used when large caliber relatively shorter barrel-like grenade launchers are required to be made. The blank is treated as an anode to create the required groove by ECM and EDM erosion.

Muzzle

- The muzzle is the front end of the barrel from which the projectile will exit.
- Precise machining of the muzzle is crucial to accuracy because it is the last point of contact between the barrel and the projectile.
- If inconsistent gaps exist between the muzzle and the projectile, escaping propellant gases may spread unevenly and deflect the projectile from its intended path.
- The muzzle can also be threaded on the outside to allow the attachment of different accessory devices.

Muzzle Blast

- The muzzle blast is the explosive expansion of propellant gases when the barrel seal previously maintained by the projectile is suddenly removed as the said projectile leaves the muzzle.
- The blast is often broken down into two components, an auditory component, and a non-auditory component.
- The auditory component, the sound of the gunshot, is important because it can cause hearing loss or give away the gun's position.
- While the non-auditory component, the overpressure wave, can cause damage to nearby items.

Problems Caused by Muzzle Blast

- Sound 140 decibels.
- The sound of a gunshot, also known as the muzzle report, may have two sources: the muzzle blast itself, and any shockwave produced by a transonic or supersonic projectile.
- Suppressors help to reduce the muzzle report of firearms by providing a larger area for the propellant gas to expand, decelerate, and cool before release.
- Overpressure wave.
- The overpressure wave from the muzzle blast of a firearm can contain a significant amount of energy because it travels at an extremely high velocity.
- Residual pressures at the muzzle can be a significant fraction of the peak chamber pressure, especially when slow-burning powders or short barrels are used.
- This energy can also be harnessed by a muzzle brake to reduce the recoil of the firearm, or by a muzzle, booster to provide energy to operate the action (Figures 5.14–5.16).

Muzzle Design

- It is necessary to keep the rifling of the muzzle end safe from accidental damage; otherwise, a damaged mouth of the rifling will result in inaccuracy because the spread of the exiting propellant gases will create uneven side pressures on the projectile. So, a concave crown is created to protect the rifling from damage and to uniformly spread the exhaust gases.

- For smoothbore barrels that fire pellet shots the spread of the pellets is controlled by incorporating chokes/constriction at the muzzle end for getting better range and accuracy. The choke can be in a range of light to extra tight, fixed, or interchangeable and screwed-in depending on the choice.

Shockwave

FIGURE 5.17 Shockwave diagram.

As the bullet crosses transonic speed and travels with supersonic speed, overpressure of 150 times or more is created as a shock wave (Figure 5.17). Unless the shooter takes care of ear protection, they may be left with a damaged ear. Appropriate muzzle devices are necessary to attenuate the shock wave pressure and suppress the muzzle flash. The constant exposure to shockwave may result in a neurological disorder in the shooters engaged in routine testing of firearms.

Factors in the Optimal Design of a Modern Gun Barrel

- Internal ballistic – consists of pressure–time, pressure–distance, velocity–time, and distance–time curves.
- Plastic stress of an internally pressurized thick-walled cylinder made of hardened steel.
- Perfectly plastic and plane stress condition.
- The yield criterion of Von Mises.
- Ballistic pressure equation as known.
- Introduction of dynamic force factor, kd.
- The firearm barrel is deemed as a vented fired pressure vessel.
- The formula for thick cylinder design will be applicable.

Design Check by Stress Analysis

Basic Equations of Thick-Walled Cylinders. See Figure 5.18

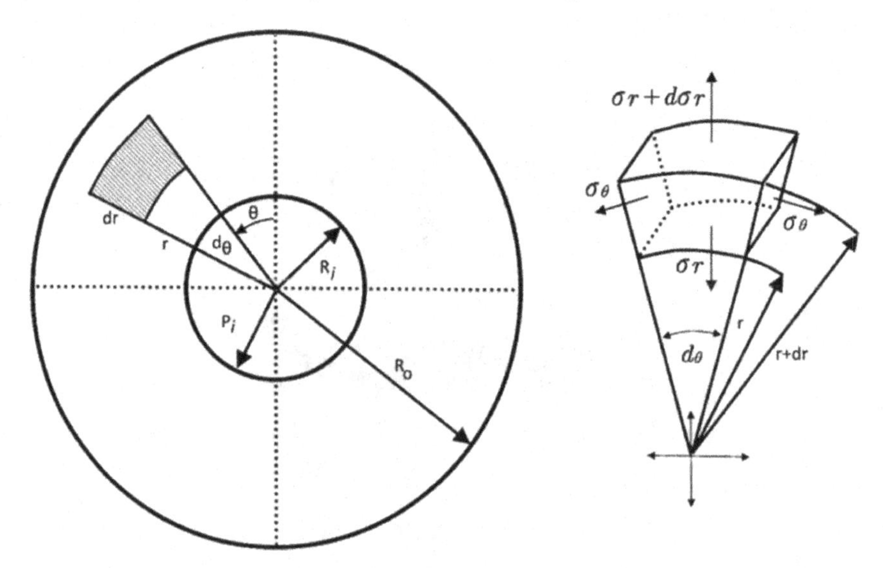

FIGURE 5.18 A thick cylinder – elemental stress diagram.

Fundamental Equations

Static condition:

$$\frac{d\sigma_r}{dr} + \frac{\sigma_r - \sigma_\theta}{r} = 0 \qquad \text{Eq 5.7}$$

where, σ_θ = hoop stress, σ_r = radial stress.

Dynamic condition:

$$\frac{d\sigma_r}{dr} + \frac{\sigma_r - \sigma_\theta}{r} = \rho\frac{dv_r}{dt} = \rho\frac{d^2r}{dt^2} \qquad \text{Eq 5.8}$$

where v_r = radial expansion velocity, $\frac{dv_r}{dt}$ = radial acceleration, and ρ = density of cylinder material.

Simplified Analysis (Assumptions for smooth bore barrel)

- The material is isotropic, homogeneous.
- The shear stress induced due to spinning of bullet by the rifling has not been considered in the analysis. Due consideration will be required for this stress for exact analysis.

- Longitudinal stresses in the cylinder wall are constant across each section of the barrel.
- The thick-walled cylinder can be considered as many thin cylinders' thickness.
- Internal or external cylinder pressure is assumed to be uniform.
- Stress amplification in the dynamic condition is taken care of by factor kd.
- The barrel is subjected to average pressure in each elemental cross-section.
- All three stresses: the radial, tangential, and longitudinal stresses are in mutually perpendicular planes. The shear stress is presumed to be absent; hence these are considered as three principal stresses.

The Solution of the Static Condition

$$\sigma_r = A - \frac{B}{r^2}, \sigma_\theta = A + \frac{B}{r^2}, \sigma_L = p_i \times \frac{\pi R_i^3}{\pi\left(R_o^2 - R_i^2\right)}$$

Gun barrel is treated as a thick cylinder subjected to an internal pressure p_i and external pressure zero. With this the boundary conditions are

At $r = R_i$, $\sigma_r = -p_i$
At $r = R_o$, $\sigma_r = 0$,

$$\sigma_r = \frac{p_i R_i^2}{R_o^2 - R_i^2} \times \left(\frac{r^2 - R_o^2}{r^2}\right) \quad \text{Eq 5.9}$$

$$\sigma_\theta = \frac{p_i R_i^2}{R_o^2 - R_i^2} \times \left(\frac{r^2 + R_o^2}{r^2}\right) \quad \text{Eq 5.10}$$

Check for Von Mises Criteria (see Figure 5.19)

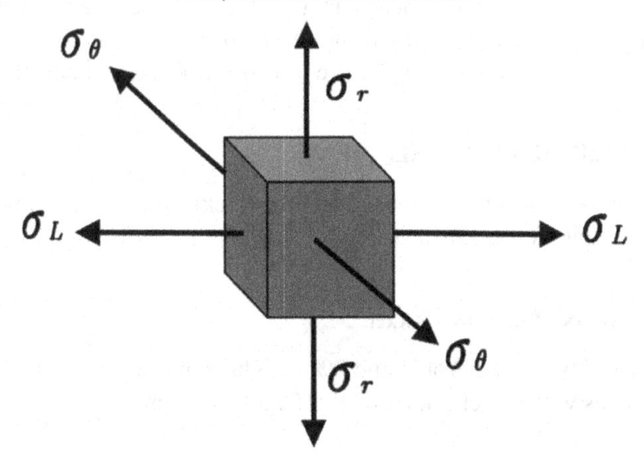

FIGURE 5.19 Principle stress on barrel element.

For applying Von Mises for design safety the following condition must be satisfied:

$$(\sigma_r - \sigma_\theta)^2 + (\sigma_\theta - \sigma_L)^2 (\sigma_L - \sigma_r)^2 = 6k^2 \le 2\sigma_y^2 \quad \text{Eq 5.11}$$

or

$$(\sigma_r^2 + \sigma_\theta^2 + \sigma_L^2) - \sigma_r\sigma_\theta - \sigma_\theta\sigma_L - \sigma_L\sigma_r = 3k^2 \le \sigma_y^2 \quad \text{Eq 5.12}$$

Gun Barrel Manufacturing: Modern Methods

- **Initial Truing**: To face the ends of the square and give the exterior a uniform roundness.
- **Deep-Hole Drilling**: Straight to within .0005 inches – is the most difficult part of the barrel-making operation. Barrels to be a button, broach, or cut-rifled are drilled several thousandths under bore size.
- **Rifling**: The most common methods of rifling a barrel include cut, broach, button rifling, Cold swaging and ECM (Electro Chemical Machining).
- **Air-Gauging**: After rifling, an air gauge – with a precision snug-fitting probe sensitive enough to detect extremely small inconsistencies in bore diameter as it travels through the barrel.
- **Straightening**: Barrels can be straightened by an experienced craftsman using a turret press.
- **Lapping**: To remove the small inconsistencies remaining in the bore, barrels are lapped as a final step. The bores of factory production barrels are rarely, if ever, lapped.
- **Stress Relief:** As a final step, virtually all aftermarket barrel manufacturers and many commercial gun makers stress-relieve finished barrels to remove internal stresses that can cause warping and wandering points of impact as the barrel heat up.

The end result, the rifled cylindrical bar, is called a barrel blank. It will undergo additional operations such as chambering, head spacing, crowning, threading, and/or further contouring and polishing before it begins service as a barrel.

BARREL MATERIAL OF SMALL ARMS

The specifications of the materials used and the thickness of the chamber and barrel walls are important factors for the safety of firearms.

Steel Specifications Used in Barrel

- Alloy steels with 1% chrome and 2% molybdenum
- Alloy steels with 3% chrome and 0.5% molybdenum

- Alloy steels with chrome, nickel, and molybdenum
- Ferritic stainless steel

Chemical Composition

The maximum content of the elements phosphorus and sulfur must be 0.025–0.035%. A high degree of steel purity is desired for elasticity with some sacrifice of machineability.

Chrome–molybdenum or vanadium steels are desired as barrel material because of their ability to retain strength even at 550°C.

Mechanical Properties

Steels for use in the manufacture of firearms, after having been heat treated, should possess well-defined mechanical properties: elasticity limit (0.2%), tensile strength, yield point, and hardness (which is proportional to the tensile strength).

Choice of Steel

The standards mentioned, as well as the methods of production of steel, account for much of their quality. The choice of steel for the manufacture of gun barrels is very complex and prescribes directly the thickness of barrel walls.

The desired elastic limit for barrel material is at least 45 N/mm^2 for smoothbore barrels and 550 N/mm^2 for rifled barrels.

A mandatory minimum wall thickness of firearms is stipulated by regulation, for example, the minimum thickness for a 12-bore shotgun at breech end is 1.1–2.0 mm tapering to 0.30–0.55 mm at the muzzle end.

Similarly, the minimum wall thickness for a rifle barrel should not be less than 1.3–2.0 mm in any section. The range of thickness depends on the strength of the steel.

Algorithm for Design Check by Energy Method

- Calculation of Detonation Products Internal Energy Per Unit of Length (E)
- Calculation of Wall Kinetic Energy Per Unit of Length (W)
- Calculation of Work of Plastic Deformation Per Unit of Length (Ef)
- Calculation of Work of Rifling Friction Force Per Unit of Length (Wr)
- Use E+W+Ef +Wr=1

Estimation of Barrel Peak Pressure from the Geometry of the Barrel and the Bullet Mass and Velocity (See Figures 5.20 and 5.21)

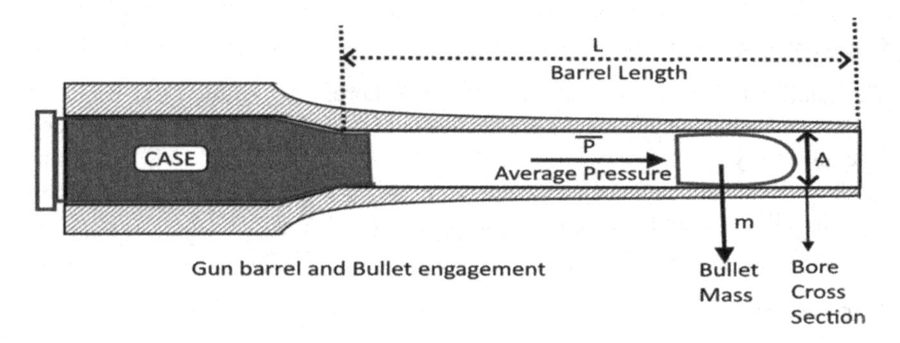

FIGURE 5.20 Gun barrel and bullet engagement.

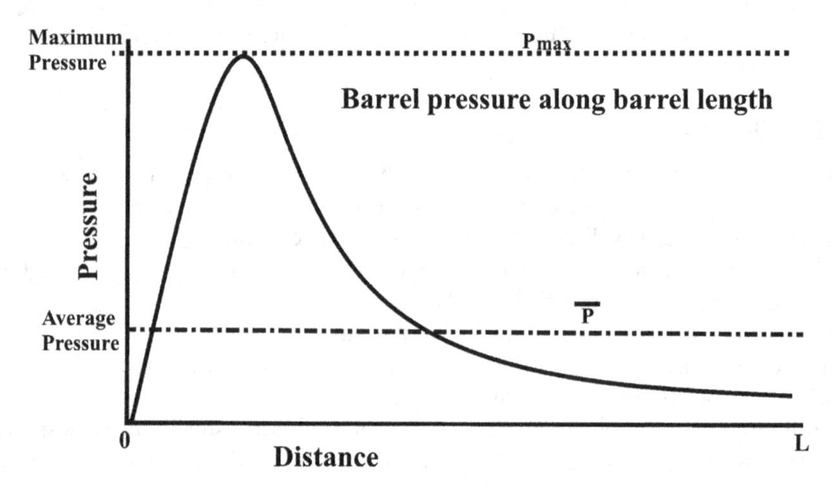

FIGURE 5.21 Barrel pressure vs distance.

The effective average force acting on the bullet (F) = average pressure × the cross-sectional area of the barrel bore = $\bar{P} \times A$

The effective work is done on the bullet = The effective average force × length of the barrel

$$= F \times L = \bar{P} \times A \times L$$

This energy is translated into bullet kinetic energy =1/2 (mv^2)

so $\bar{P} \times A \times L = \frac{1}{2} mv^2$, hence $\bar{P} = \frac{mv^2}{2AL}$

The propellant energy has to overcome the barrel friction, and besides has to supply energy for elastic expansion of the barrel and acceleration of the bullet, which is of the order of around several thousands of 'g' (acceleration due to gravity). As the gas expands, the peak pressure developed in the barrel is inversely proportional to the square of the bore and directly proportional to the projectile mass. This results in an instantaneous peak pressure which is 2.25 to 3.70 times more than the average pressure depending on the caliber of the ammunition and grain of the projectile.

The peak pressure will be directly proportional to the projectile mass and inversely proportional to the square of the diameter for a given grain of propellant. It means the higher the peak pressure, the higher the sectional density of the projectile. The nature of the variation in the correction factor is illustrated in Figure 5.22.

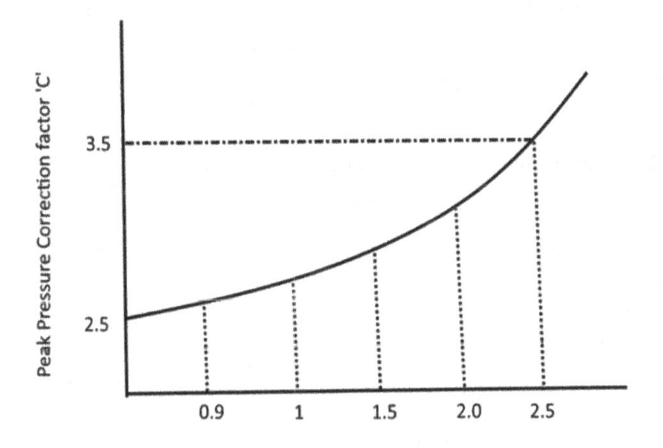

FIGURE 5.22 Peak pressure correction factor vs sectional density of projectile.

Therefore for a realistic estimate, a correction factor 'C' within the range 2.25–3.70 has to be applied for the estimation of peak pressure. Thus we can write

$$P_{max} = C \times \frac{mv^2}{2AL} = C \times \frac{W}{g} \times \frac{v^2}{2AL} \qquad \text{Eq 5.13}$$

It is noteworthy to point out that it is the effective average gas pressure impulse that is responsible for the automatic operation of the gas-operated rifle. Therefore, the location of the gas trapping point on the barrel is decided greatly on the effective average pressure. So, one should not be surprised to see a gas trapping hole almost near the chamber for automatic shotguns whose average pressure is considerably low. (Figures 5.23–5.26).

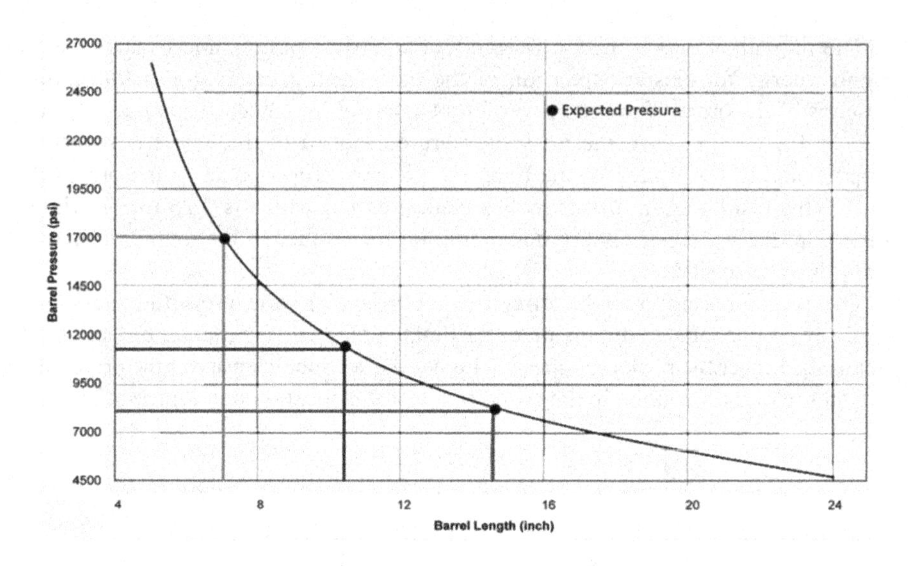

FIGURE 5.23 Barrel length vs bore pressure.

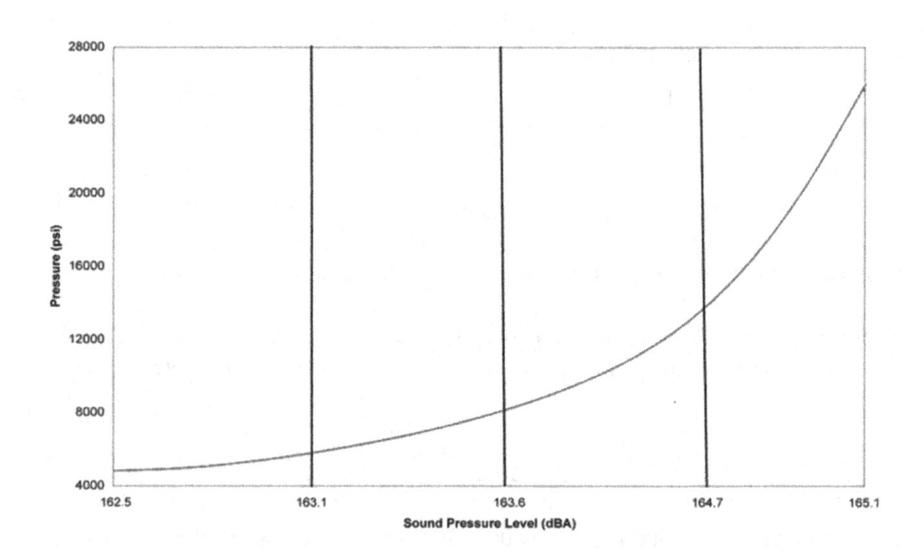

FIGURE 5.24 Pressure vs sound level (dB). (The graph has an exponential trend.)

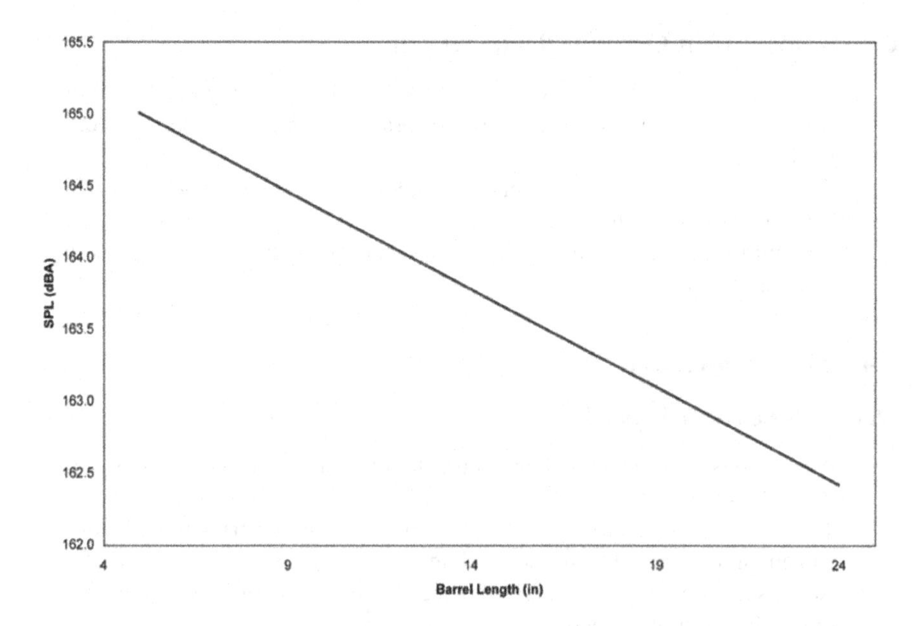

FIGURE 5.25 Sound pressure level (SPL) vs barrel length.

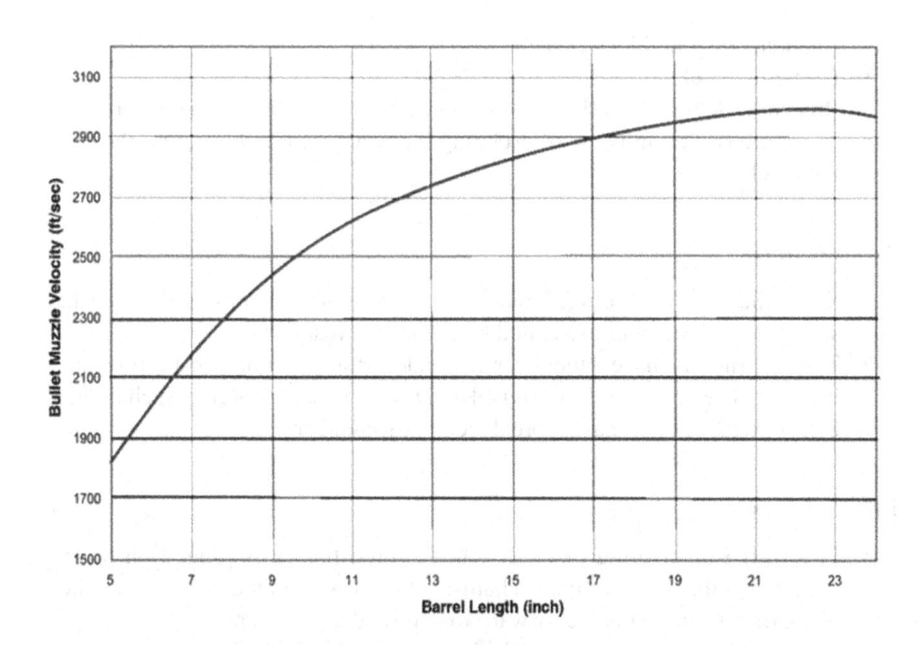

FIGURE 5.26 Bullet muzzle velocity vs barrel length.

Conclusions about Optimum Barrel Length

1) Pressure in the bore increases exponentially with decreasing barrel length.
2) Non-suppressed sound pressure level increases exponentially with decreasing barrel length.
3) Pressure in a suppressor's entrance chamber increases proportionately with expected bore pressure.
4) Maximum velocity is attained in an approximately 21-inch barrel of 5.56 mm caliber.

5.4 BLOCK BREECH

Design Principle of Block Breech

- An action is a physical mechanism that manipulates cartridges and/or seals the breech.
- The term is also used to describe the method in which cartridges are loaded, locked, and extracted from the mechanism.
- Actions are generally categorized by the type of mechanism used; the block breech is a part of the action.

Types of Actions (Single-Shot-Manual action)

- Break action – no block breech.
- Dropping block.
- Breechblock lowers itself or "drops" into the receiver for opening the breech.
- There are two main types of dropping block: the falling block and the tilting block.

Break-action

- When the barrels are hinged and can be broken open to present the breech, such types of weapons are called break-action weapons.
- The firearms can have either over and under or side by side two-barrel combination. They can also be designed as a mixed rifle or shotgun. Such multi-barrel firearms often have a break-action mechanism.

Tilting Block

- In a tilting or pivoting block action, the breechblock is hinged on a pin mounted at the rear. In this mechanism, a lever is operated, after which the block tilts down and move forward to expose the chamber.
- The well-known pivoting block designs are the Peabody, the Peabody-Martini, and Ballard actions.

Falling Block

- A falling-block action (also known as a sliding-block action) is a single-shot firearm action in which a solid metal breechblock slides vertically in grooves cut into the breech of the weapon and actuated by a lever.
- Examples of firearms using the falling block action are the Sharps rifle and Ruger No. 1.

Rolling Block

- In a rolling block action, the breechblock takes the form of a part-cylinder, with a pivot pin through its axis. The block is opened or closed simply by rotating the block around its axis.
- Rolling blocks are most often associated with firearms made by Remington in the later 19th century.
- In the Remington action, the hammer serves to lock the breech closed at the moment of firing, and the block in turn prevents the hammer from falling with the breech open (Figures 5.18–5.26).

5.5 MUZZLE ATTACHMENTS

Muzzle Brake, Compensator, Flash Hider, Silencer, and Sights.

Muzzle Brake

FIGURE 5.27 Muzzle brake.

A muzzle brake is a device connected at the end of a barrel that redirects a portion of the propelled gasses to reduce muzzle rise, linear movement, and recoil (Figure 5.27). Muzzle brakes direct gasses up, to the sides, or backward to achieve the prior goals, typically making the brake extremely loud. A good way to understand muzzle brakes is to look at Newton's third law: for every action, there is an equal and opposite reaction. So when a brake redirects hot, fast-moving gases upwards, the muzzle is then pushed down with the same force as the gases leaving. This is how muzzle brakes can control muzzle flip, and perceived recoil.

Flash Hider (or Flash Suppressor)

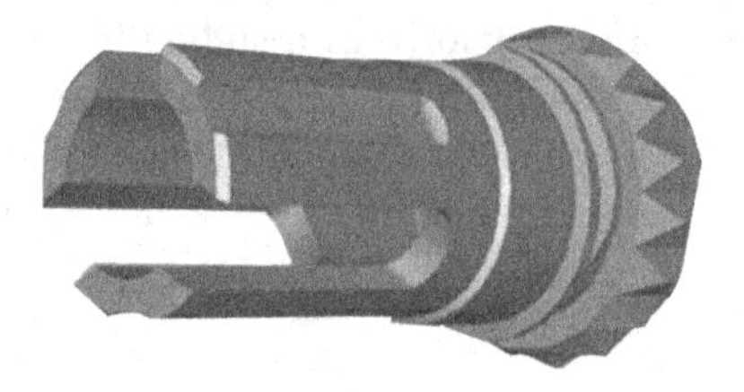

FIGURE 5.28 Flash hider.

Muzzle flash is not due to unburned powder (Figure 5.28). Flash is extremely hot gases (usually around 3600°C) expanding rapidly when meeting cooler ambient air. These hot gases leave the end of the barrel and create a sort of bubble. The "flash" that we see is this gas bubble becoming incandescent and briefly crossing the visible spectrum. In simpler terms, hot gases make a bright light. To negate this burst of visible light, flash hiders create turbulence. By blending the hot gases with cooler ambient air in a disruptive pattern, there can no longer be a large bubble to glow, and the gas is dissipated in smaller sections.

Compensator

FIGURE 5.29 Compensator.

A compensator, much like a muzzle brake, is designed to manipulate expelled gases to achieve less recoil and muzzle movement (Figure 5.29). Whereas the brake tends to aid greatly in reduced recoil and only slightly in lessened muzzle movement, the compensator does the exact opposite. Reducing recoil is a compensator's secondary goal, with less muzzle rise or linear movement as its primary objective.

Silencer and Suppressor

Silencer

A silencer is a muzzle device that reduces the acoustic intensity of the muzzle report and eliminates muzzle flash when a gun is fired, by manipulating the speed and pressure of the gas ejection from the muzzle and hence suppressing the muzzle blast. Similar to a muzzle brake, a silencer can also be detachable, being mounted separately onto the muzzle.

Silencer vs Suppressor

A silencer is mainly for reducing the sound, while a suppressor is more for eliminating muzzle flash and it also reduces some of the sounds.

Baffles and Spacers

Baffles are usually circular metal dividers that separate the expansion chambers. Each baffle has a hole in its center to permit the passage of the bullet through the silencer and toward the target.

Types of Baffled Silencers/Suppressors

1. **Monolithic Baffles**:

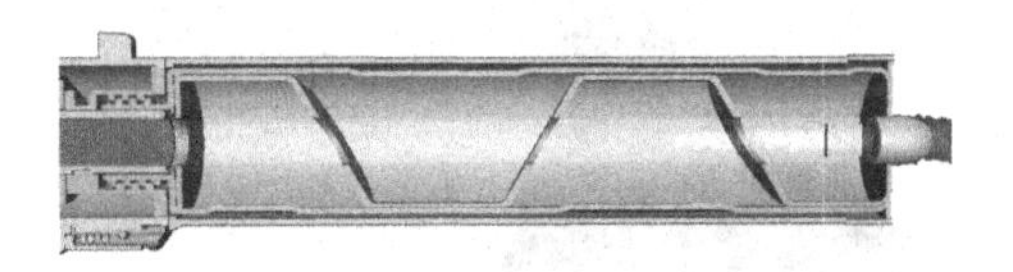
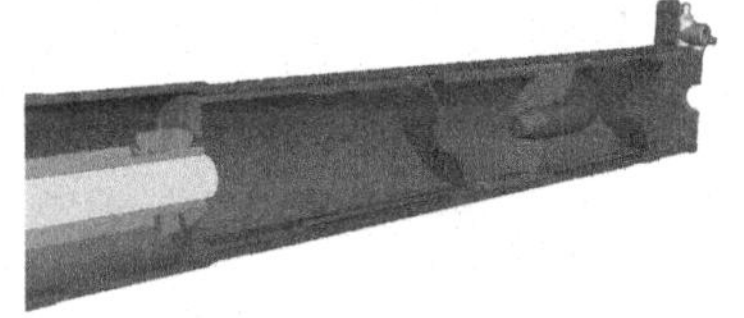

FIGURE 5.30 Monolithic baffle suppressor.

The monolithic core is one solid piece of metal that has had sections machined out of it or sometimes created through casting. The stacked baffles are a series of individually created baffles that are stacked together to create the core (Figure 5.30).

2. **Stacked Baffles**:

FIGURE 5.31 Stacked baffle suppressor.

The stacked baffles require a little more precision, supplies, and labor. The baffles must be aligned straight to avoid baffle strikes; also, the baffles must almost lock into place together to avoid them from moving and messing up the alignment (Figure 5.31).

Firearm Sight Attachments

While using different types of guns, it should be possible to accurately aim the gun at the target. To do that, some sort of sighting device is needed on the gun.

The most basic sights have one point on the end of the barrel and one closer to the rear of the gun that are aligned with each other and the target. Others are more advanced.

Different Types of Weapon Sights

- Iron/Open Sight
- Telescopic Sight
- Red Dot Sight
- Reflex Sight
- Laser Sight

Iron/Open Sight

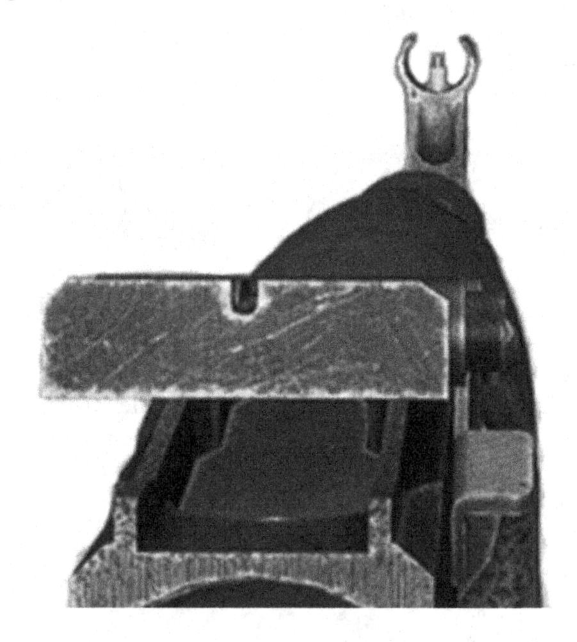

FIGURE 5.32 Open sight.

Iron sights are designed in two different ways: **open** and **aperture**

Open sights shotguns, handguns, Assault, and hunting rifles are provided with open sights. They're as simple and just have a notch for the rear sight and a post in the front. These are imprecise sights. They're low-cost, durable sights (Figure 5.32).

Aperture sights

FIGURE 5.33 Aperture sight.

When a rifle is made ready for battle zero and maximum point-blank range as is required in target and combat rifle aperture sight is the best low-cost choice (Figure 5.33). These are used generally in target shooting rifles and combat rifles. These work well when the target is close to the shooter. They consist of an aperture (like a camera) in the rear, and a sight that is commonly a post in the front or another aperture (called a peep sight, generally used on target rifles). Aperture sights are more precise than open sights, and often faster as well, but can occlude the target and are more expensive and less durable. **Shotguns** typically just have **bead sights.** Usually, just a bead is fixed at the end of the barrel; it just allows a way to index where the shooter pointing.

Telescopic Sights

FIGURE 5.34 Telescopic sight.

Telescopic sights have a graphic image pattern reticle placed in a position that gives the most optimal aim (Figure 5.34). These are used on firearms, as well as surveying equipment and some bigger telescopes.

For targets that are either very small or very far away, telescopic sights are used. Telescopic sights offer magnification of the target. Magnifications between 2x and 24x are common. While some scopes have a fixed power (2x, 4x ...) many are variable. Variable scopes generally have a limited range of magnification. When looking at scope specs, you'll sometimes see them listed as a 6-18 x 35. This means that the front lens (the "objective" lens) is 35 mm in diameter. Larger lenses allow more light to pass, but the quality of the optics decides the quality of the image through the scope.

When selecting a scope, you have to select one having the appropriate "eye relief". The eye relief is defined as the distance between your eye where you can get a clear image of the reticle (crosshairs in the scope that you center on the target) and the eyepiece (ocular lens) of the scope. For rifles, the eye relief is comparatively short because your eye will be almost touching the eyepiece. For other applications (like for use on a pistol, or what's referred to as a "scout rifle"), the eye relief must be greater because the scope won't be near your eye.

Red Dot Sight

FIGURE 5.35 Red dot sight.

A red dot sight is a type of non-magnifying reflector (or reflex) sight for firearms, and other devices used for aiming, which gives the user a point of aim in the form of an illuminated red dot (Figure 5.35). In a standard design, a red light-emitting diode (LED) is used at the focus of collimating optics which produces a dot-style illuminated reticle that remains in alignment with the weapon to which the sight is attached regardless of eye position (almost parallax free). They are considered to be fast-acquisition and easy-to-use gun sights for target shooting several types of sights may be referred to as "red dot" sights. For some scopes, there may be a red dot in addition to the reticle. Generally, "holographic" and "reflex" sights are known as red dot sights.

Reflex Sight

FIGURE 5.36 Reflex sight.

Reflector sights will be a non-magnifying sight that shows an aiming point over the target (Figure 5.36). An Irish optical designer conceived the idea of this sight in 1900 and further development progressed from there. Reflex sights make use of a LED to project a dot (or reticle or some sort) on the image seen in the sight. Some reflex sights have only 1 type of reticle. The one below has four different reticles. The problem with this is that you have to re-zero the sight if the lever is moved (accidentally or intentionally to change the reticle). This device has three levels (brightness) settings and two settings for night-vision. The night-vision settings are essentially invisible to the naked eye and are intended to be used with a night-vision scope that would be mounted behind the reflex scope.

Laser Sight

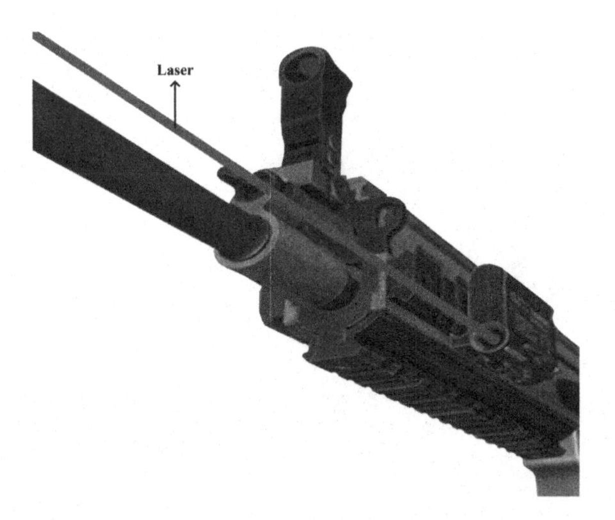

FIGURE 5.37 Laser sight.

The laser in most firearms is used to enhance the aiming ability of the weapon systems (Figure 5.37). A *laser sight* is a small, visible-light laser placed on a rifle, as shown in the adjoining figure, and aligned to emit a beam parallel to the barrel axis. This can also be used in a handgun. Since a laser beam diverges less, the laser light is projected as a small spot at long distances, too; the user places the spot on the desired target and the barrel of the gun gets aligned. However, the automatic correction for bullet drop, windage, does not happen. A red laser diode is used in most of the laser sights. It is also possible to use an infrared diode to produce a dot invisible to the naked eye, which is detectable with night-vision devices.

BIBLIOGRAPHY

Bill Holmes, *Home Workshop Prototype Firearms: How to Design, Build, and Sell Your Own Small Arms*, Paladin Press Book, USA, 1994.

Brian J. Heard, *Handbook of Firearms and Ballistics: Examining and Interpreting Forensic Evidence (Developments in Forensic Science)*, Wiley-Blackwell, UK, 2008.

Oliver B. Hamilton, *Advanced Gunsmithing by W.F. Vickery*, Kingsport Press Inc, Tennessee, 1940.

6 Quality of Design

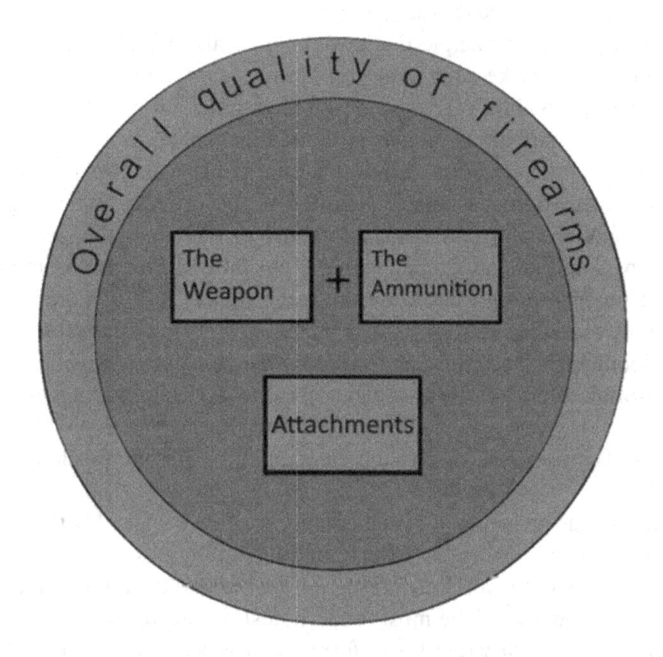

FIGURE 6.1 Overall quality of firearms.

Overall quality of firearms (Figure 6.1).

THE GENERAL DESIGN PHILOSOPHY

- **Modern design** – Modern design is a design based on the principles of modernism, such as "form follows function" and "less is more".
- **Universalism** – The belief that there is a universal set of rules that all designers should follow.
- **Internationalism** – Modern design embraces internationalism whereby designs are the same for all countries as opposed to being crafted to reflect the local culture. For small arms, advanced countries have their designs that are unique in nature.
- **Science and technology** – Modern design aligns with science and technology as opposed to culture and nature.

DOI: 10.1201/9781003199397-6

- **Attractiveness principle** – The principle that a design cannot be made visually appealing for everyone because individuals have different perceptions of attractiveness. Modern design doesn't try to visually impress but instead takes the conservative route of trying to be visually inoffensive.
- **Ornament and decoration** – Modern design and architecture largely rejecting decoration and ornamentation.
- **Scalability** – The principle that designs be scalable such that they can be manufactured, marketed, and delivered to very large markets.
- **Industrial materials** – The use of the most practical material for a design including considerations such as cost and mechanical properties. This usually results in the selection of industrial materials such as plastic and steel.
- **Dematerialization** – Dematerialization is the process of making things lighter, smaller, and less material-intensive, including the possibility of removing materials altogether with virtual things. This is a common feature of modern design based on the principle of "less is more". It means achieving maximum attributes and functions by using minimum materials.
- **Open concept** – Design is carried out by the concept of "less is more".
- **Functional color** – Colors are never used to decorate but are selected for their function, such as visual contrast or brightness.
- **Unity** – The principle that visual elements look like they belong together as a whole.
- **Balance** – The principle of visual balance is often achieved with symmetry.
- **Order** – Modernism views science and technology as a force that is destined to reorder the world according to rational and universal principles.
- **Mechanical forms** – The most modern design uses mechanical geometrical forms, particularly rectilinear forms such as grids. These are considered more orderly than organic shapes, such as a free-flowing line.
- **Standardization** – The development of design standards at the level of society, industry, organization, and team that reduces design to the application of rules.
- **Modularity** – Breaking things into modules based on function.
- **Repetition** – Modern design sees nothing wrong with relentless repetition and this may be favored as a symbol of order. This results in a system of small arms (Figure 6.1).

FUNCTIONALITY AND FEATURES OF SMALL ARMS BASED ON COMBAT ROLES

- Rifles
- Carbine
- Sniper rifle
- Machine guns
- Pistols
- Revolvers
- Grenade launchers
- Shotguns

DIFFERENCES BETWEEN RIFLES, SHOTGUNS, AND HANDGUNS

The main differences are in their barrels, their targets, and the types of ammunition.

Rifle

The rifle has a long barrel with thick walls to withstand high pressures and rifling to impart the spin to increase the accuracy and range. Each rifle fires only a specific caliber of ammunition. The configurations are available in bolt action, semi-auto, and fully auto mode. The design of their trigger mechanism is quite complex. In action, the rifles may use any of the principles of gas operation and derived recoil operation.

Handgun (Pistols/Revolver)

The handgun has a short-rifled barrel with thick walls to withstand high pressures. But it causes considerable muzzle jump during firing. It can be used for shooting targets at a much smaller range. Similar to rifles, the bore of a handgun barrel is usually made for only one particular type of low peak pressure ammunition. The revolver normally consists of a cylinder with multiple chambers and is operated in either SA/DA mode. The inevitable gap between the cylinder and barrel of the cylinder always leaves a gunshot residue (GSR) on firing. The silencers are useless in revolvers. Pistols are semi-automatic and may be fully automatic, and hence have a more complex design. They can be equipped with suitable silencers.

Difference between a Rifle and Carbine

In operation principles, the carbines and rifles are similar. With the only difference that with low powered ammunition, a carbine can work in blowback principle, which may result in a simpler firing mechanism. The essential difference between the two is in the length of their barrel. A barrel length of lesser than 20 inches is considered a carbine class. Like rifles, they can be used in the close-quarter battle.

Difference between an Assault Rifle and Sub-machine Gun

FIGURE 6.2 Sub-machine gun and ammo.

Sub-machine guns are machine guns that fire pistol-sized ammo and meant to be a compact version of the machine gun. The mechanisms can be simple or complex depending on the choice of the principle of automation, whether operating on straight blowback or other principles.

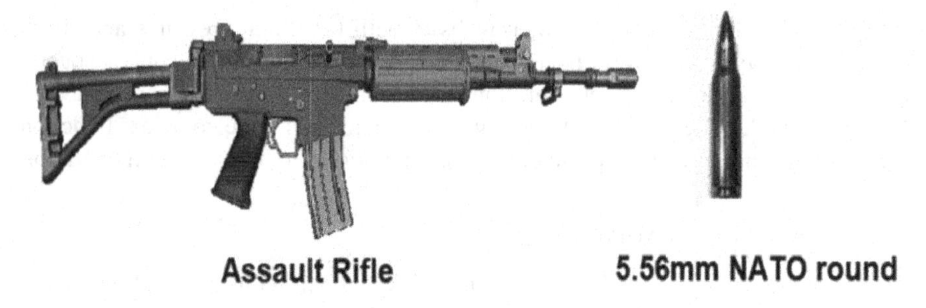

FIGURE 6.3 Assault rifle and ammo.

The essential features of an assault rifle are that they must have a firing mechanism for fully automatic and burst fire and they must be fed by a box-type magazine that holds intermediate power cartridges. They are compact, while at the same time having more ranges and power than a sub-machine gun. Currently, these are available in calibers of 5.56–7.62 mm (Figures 6.2 and 6.3).

Difference between an Assault Rifle and Machine Gun

- M-16, AK-47, and Tavor X-95 are examples of modern assault rifles and they hold a magazine containing about 30 rounds. Machine guns are automatic weapons required for delivering a high volume of fire. As such, these machines have a large drum-type magazine to hold 100 rounds or more; alternatively they are designed with complex mechanisms to feed and fire belted ammunition.
- These are designed to fire in an open bolt position to avoid the problem of ammunition cook-off. Normally, these are provided with a replaceable and extra barrel. These are mounted on a vehicle or platform and can have their bipods and tripods as additional accessories SAWs, M-60s, and MG-43 are examples of well-known machine guns. Machine guns do not have a single-shot option and are always fired in the automatic mode.

Difference between Pistol and Revolver

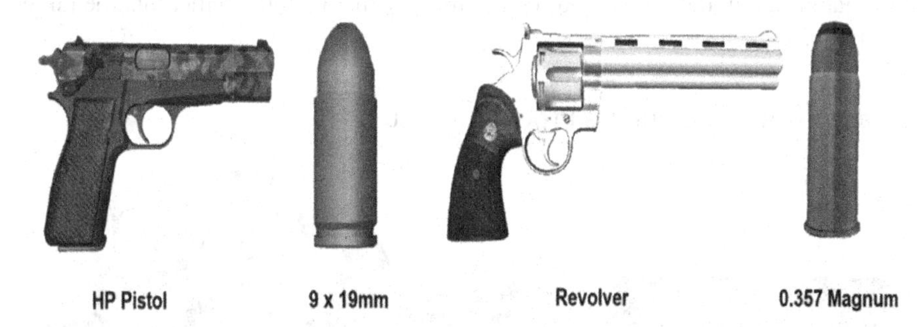

FIGURE 6.4 Pistol vs revolver and their ammo.

- Pistols are semi-automatic machines. Though pistols can be designed using any of the basic automation principles, modern pistols operate on a short recoil principle and they are normally fed by a magazine fitted into the pistol grip. They have a slide that carries the barrel and transfers the recoil energy for generating automation. The pistol can have one or more replaceable barrel, each containing a specific chamber. The cartridge headspace of the pistol starts from the mouth of its casing to its back. The cartridges are normally rimless.
- The revolvers are devices that carry a revolving unit containing several chambers which can be loaded either by swinging the cylinder sideways or by a vertical breakup. They also have a rifled barrel, but they are provided with clear gaps with the chamber for free rotation of the cylinder. The revolver uses more powerful cartridges than the pistol – the cartridges are rimmed and the headspace is on the front face of the rim (Figure 6.4).

Performance

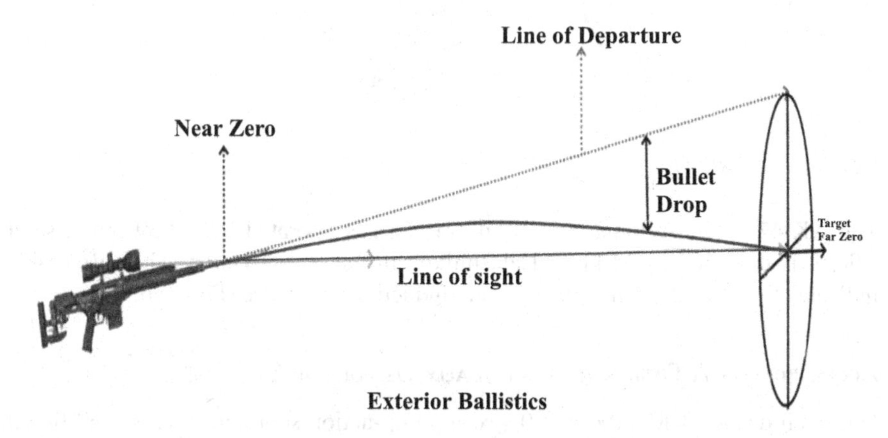

FIGURE 6.5 Exterior ballistics.

The performance of a firearm is decided by the remaining energy of the bullet, its terminal effect on the target, and the accuracy with which the bullet hits the target (Figure 6.5).

Usability – A Design that is Pleasing to Use

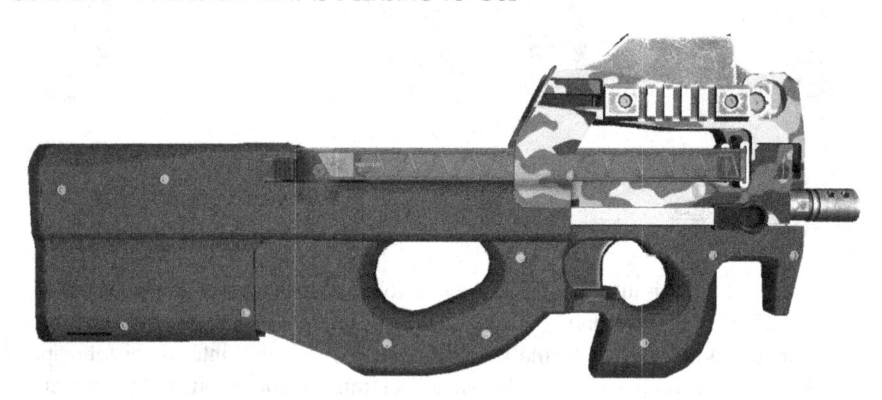

FIGURE 6.6 FN P90.

It is a Bullpup design that has an integrated reflex sight and it is fully ambidextrous to control. It fires 5.7 × 28 mm ammunition which imparts a high velocity to the projectile. As an additional feature, it has interchangeable visible or infrared laser and tritium light sources. (Figure 6.6).

FIGURE 6.7 AK 47.

The AK 47 – The Avtomat Kalashnikova is a gas-operated 7.62 × 39 mm assault rifle. It is designed by Mikhail Kalashnikov in the Soviet Union. The suffix "47" indicates the year of its manufacture and induction in service (Figure 6.7).

Accessibility – A Design that is Equally Useful for Everyone

One such design is Mossberg 500 spx, a pump-action shotgun. This is used by all types of users, including the military, civilians, and other security forces (Figure 6.8).

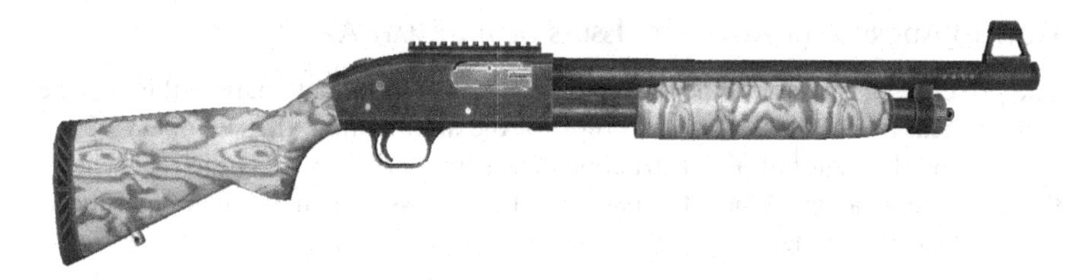

FIGURE 6.8 Mossberg 500 SPX, pump-action shot gun.

Reliability

Reliable designs are those that survive and work fine in real-world conditions over its lifecycle. The probability of failure is very low and the quality of design is such that it functions whenever it is needed. And in the case of small arms, the failure that occurs in the form of stoppages, breakage of parts is acceptable only over firing few thousands of rounds.

Small Arms Design Environment

1. Very high-pressure ammunition (typical pressure for rifle and machine guns: 300–350 MPa; for pistol and revolver: 150–200 MPa)
   The average working pressure for grenade launchers, and Shotguns: 90–140 MPa.
2. Low ballistic efficiency (25%), producing very high heat
3. Fouling by propellant gases
4. The bore is beset with the problem of lead deposits
5. Highly constrained movement of trigger mechanism parts
6. Violent automatic action and associated vibration
7. The requirement of form to function
8. Working stress nearly yield point strength
9. Climate and terrains can be extremely hot/cold and hostile

Predictability

A design is built around certain predetermined performance parameters to be delivered during the service of the product. Therefore, a predictable design is a design that work as people expect in terms of a specified parameter.

A product should be simple to use and maintain to be considered as a good design. However, if a user interface requires complex training to use and maintain the product, it may be poorly designed.

A weapon is expected to work without fail in the battlefield condition. But if it requires frequent lubrication and replacement of parts, it is a poor design.

How to Address the Reliability Issues of the Small Arms?

The principle of operation, the design of gas flow path for automation, the arrangement of heat transfer, the size and power of the ammunition, the chemistry of propellant, and the material of construction of various parts, all affect the reliability of the designed weapon. Hence, the issue can be addressed by making an appropriate choice of the following.

Choice of the Principle of Operation

a) Gas operation (DI, LS, SS)
b) Blowback (SBB, DBB)
c) Recoil operation (LR, SR)
d) Manual operation

Choice of the Material of Construction

a) Metal, steel vs brass
b) Plastics vs composites
c) Composites

Design Choice

a) Minimal components
b) The robust design means a high fault tolerance level
c) Form to function
d) Less is more, meaning maximizing function with minimum components
e) Interchangeable vs customized
f) Modular vs non-modular

Twist rate – Rifling is specified by its twist rate, which indicates the distance the rifling takes to complete one full revolution. The twist rate 1 turn in 12 inches (1:12 inches) implies one full revolution by the rifling will be completed after traveling 12 inches. A high twist rate imposes higher stress on the barrel resulting in a high rate of wear and lower barrel life.

Polygonal Rifling – Polygonal barrels tend to have longer service lives compared to conventionally rifled barrels because of the reduction of the sharp edges of the land.

Gain-twist rifling (Progressive Rifling) – Gain-twist rifling, also called progressive rifling, begins with very little change in the projectile's angular momentum during the first few inches of bullet travel after ignition during the transition from chamber to throat.

Consistency – Consistency is the result of a user interface/ manufacturing platform with the same controls on every part.

Manufacturing Process

The manufacturing process is a very influential decision for imparting physical, mechanical, and chemical properties of parts and components and hence affects the quality of a product. For example, a metal part produced through conventional machining, casting, or powder metallurgy process. The quality of the parts produced through different methods will be distinctly different; hence the choice of the manufacturing process from the point of end-use is of utmost importance to assure the desired quality requirement.

For example, the choice of manufacturing processes for barrel can be varied by choosing different rifling and surface treatment processes. This will also depend on the type of small arms that is under consideration for production. Each of the barrel for sniper/assault rifle/handguns/shotguns will go through unique processes.

Rifling Processes

- Indexing and cut process – Suitable for low volume and high precision, small caliber barrel
- Cold swaging or/and button dyeing – Suitable for high volume of barrel production
- Broaching or electrochemical machining – Suitable for large bore, low batch production

Surface Treatment

The qualitative difference in the surface properties may be obtained by choosing alternatives as under:

- Electroplating or electroless plating – To impart surface hardness and corrosion resistance properties, the electroless plating can be applied to non-metal parts also.
- Parkerizing or cerakoting – To impart properties of corrosion resistance; however, cerakoted parts exhibit excellent properties of abrasion as well as heat resistance.
- QPQ or carbonitriding – Both the processes aim at imparting surface hardness to the metal part. But the QPQ process gives high polishes to the metal surface in between the treatment and this imparts a superior tribological property to the metal surface. Therefore, if trigger mechanism parts and the bolts are expected to function flawlessly if treated with the QPQ process.

Stability – Error-free Designs

There shall be a built-in ability to continue in a reasonable way when an error occurs.

For example, a machine gun shall not falter and suddenly halt and catch fire every time an error occurs.

An assault rifle will fire and reload a cartridge if the magazine has ammunition.

Safety and Security

The design of the device must have an arrangement for safety and security. For example, the change lever system applies for mechanical safety to avoid accidental fires and reduce human error.

The weapon shall not cause an unintentional fire.

It shall not fire by an accidental fall.

Upgradability/Reusability

The design must be reusable and extensible. For example, the modular design of rifles in which a weapon allows the barrel to be upgraded to work like a machine gun as opposed to requiring a completely new device when you need more firepower.

Emotional Durability and Experience

There are intangible elements of quality, such as a business tool that is as engaging as a game and a design that people value at an emotional level which they don't easily throw it out. AK 47 rifle or MP5 carbine also shares the same type of emotional attachment; if any of their parts fails, then the user tries to repair it so that it can be reused.

Refinement

The overall sophistication and elegance of design must have a necessary platform for the attachment of auxiliary devices, for example, optical scopes, grenade launchers, and fitment of muzzle devices.

Quality of Design of Firearms

Firearms are used by several sections of people for a range of activities that encompass recreation and self-defense by the civilian, assault, and deterrence by the security forces – strategic killing and offense by the military and the militant groups.

The firearms are used for several situations and are expected to function in all environments and terrain reliably without endangering the safety of the users and bystanders. In parametric terms, the different class of weapons has their specification parameters such as muzzle velocity, range, rate of fire and recoil, etc., which are made available by the manufacturer. However, such parametric attributes do not necessarily guarantee the quality of the firearm.

The phrase "Quality is the customer's delight" is absurdly inappropriate in the context of a firearm. The aesthetics and ergonomics very often take an overriding call over the inner core quality.

However, for the pragmatic men in uniform, the proven philosophy as depicted in the diagram –small arms design philosophy – quality by design will be gainful and advantageous for the selection of the weapon.

Further Points of Importance That Are to Be Borne in Mind in the Design of Firearms

The firearm is a modified form of I.C. engine where the power stroke is performed by the projectile.

It is subject to all the constraints governed by the laws of thermodynamics and demands efficient thermal management while designing automatic weapons for volley-fires.

- The mechanisms of a firearm are under extreme stress and wear while in operation. The toll on the weapon by friction can be hardly controlled by conventional lubrication.
  The complexities of the parts of the mechanism are not amenable to cost-effective mass production through conventional metal cutting.
- The figure and the forms of the user-friendly firearms can be achieved only through techniques like MIM (metal injection molding), PM (powder metallurgy) for metallic parts, and injection molding of high-grade polymers.
- To ensure the quality of design, it becomes necessary to stipulate special processes and materials in the design document and process layout. The design and the process audit become an important management function to produce firearms with consistently high quality. To name a few processes like button dyeing/cut machining, MIM/investment casting, EN/electroplating austempering/martemparing carbonitriding/QPQ, and painting/cerakoting are to be specifically stipulated in the process layout.
- A firearm in stand-alone mode is not sufficient for meeting the quality need. The interior and the exterior ballistics of the bullet greatly design the end performance. The specified performances are therefore to be interpreted in conjunction with the specific ammunition. The universality of a weapon concerning the ammunition would have been an ideal goal; however, this is not possible because of the individualistic character of each firearm, and the presence of a broad spectrum of caliber and size of the weapon.
- The safety of the weapon is addressed by various established proof standards and standard specification of material for stressed parts. Reliability issues are covered with a full proof design of kinematics and proper specifications of tolerances and physical properties of the parts and the components. The choice from the various principles of automation becomes very relevant in ensuring trouble-free operations and management of heat-related problems.

- Recoil forces and energy dictate the perceived quality of the weapon. And this is very important as we all know, perception about anything is more important than the reality to every one of us. So, variations among shapes, colors, sizes, and the use of contemporary materials speak a lot in spelling out the quality of the weapon. So, no less attention shall be paid to the exterior appearance of the firearm. The potential of computer graphics should be fully utilized to make a virtual design before the actual production, to meet the ever-increasing users' expectations (Figure 6.9).

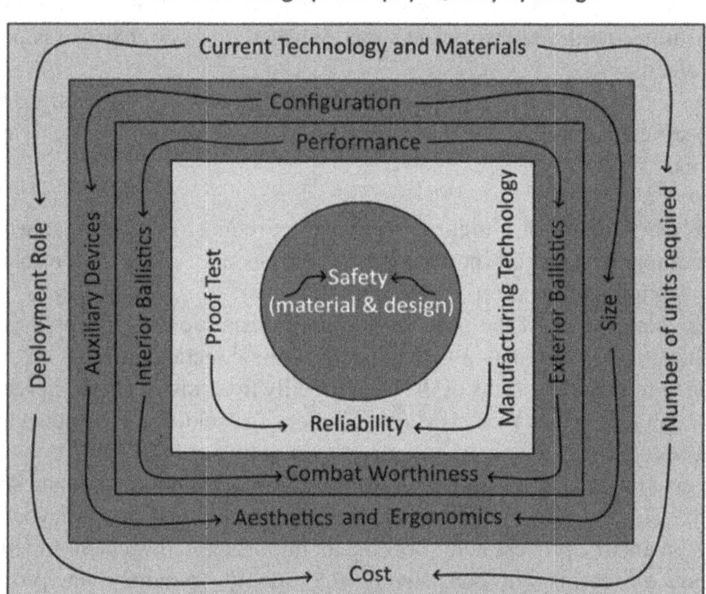

FIGURE 6.9 Small arms design philosophy – quality by design.

BIBLIOGRAPHY

J. Juran, and A. Blanton Godfrey, *Juran's Quality Handbook - 5th Edition*, The Mc Graw Hill Companies, USA, 1998.

T. Pyzdek, and P. Keller, *The Handbook for Quality Management: A Complete Guide to Operational Excellence*, McGraw Hill Education, USA, 2013.

7 Special Processes in Small Arms Manufacturing

THE NECESSITY OF SPECIAL PROCESSES

The manufacture of small arms is a highly competitive global business. It requires parts of extremely complex shapes and materials with mechanical properties. Anyone a little familiar with manufacturing small arms knows that to produce finished small arms components by conventional forging and machining routes, the material utilization is no better than 60%, and there are cases in which the material utilization is as low as 10%. This results in a lot of waste as turnings, boring, kneeling, and swarf. Low utilization of material is an added premium to the cost and the disposal of this waste is a burden to the management, being an environmental hazard.

The exclusivity high performance, aesthetics, bio-compatibility, marketability, and green production have become buzzwords in today's industrial scenario. The small arms cannot remain as an exception. To win over a place in the competition, many firearm companies in the world are adopting special techniques of production such as cold forging, EN plating, ceramic coating, MIM, etc., in their production process.

This chapter aims to introduce the readers to a basic concept of some of these special processes to appreciate the subject.

MANUFACTURING APPROACHES FOR INTRICATE/COMPLEX SHAPES

I. (Bottom-up) – The starting materials are in elemental shape and they are added up to build up the final product either by assisted assembly or self-assembly. The process is very fast and proceeds exponentially. The material utilization is 95–100 % in this production technique for a complex part.

II. (Top-down) – The starting material is a big mass and the final product is produced by deduction/removal of material from the parent block. The material utilization may be as low as 10–20% in this manufacturing technique of complex shape.

Focus – Choice of approach primarily focuses on material removal ratio/material utilization in the production of parts.

DOI: 10.1201/9781003199397-7

MANUFACTURING TECHNOLOGY

- CNC machining
- Injection molding
- PM – Powder metallurgy
- MIM
- Investment Casting

Metal Injection Molding and Powder Metallurgy

Powder Metallurgy Basics

- Begin the process with blending or mixing specific types of metal powders and lubricants/binder to hold the powder base in a semi-rigid state, which assures segregation-less blend, even when those granules exhibit a natural inclination for separating.
- Place the mix in a die to create a certain shape.
- Compact the metal powder tightly using a press.
- To form metallurgical bonds in the metal powder, sinter the compressed part in a furnace (Figure 7.1).

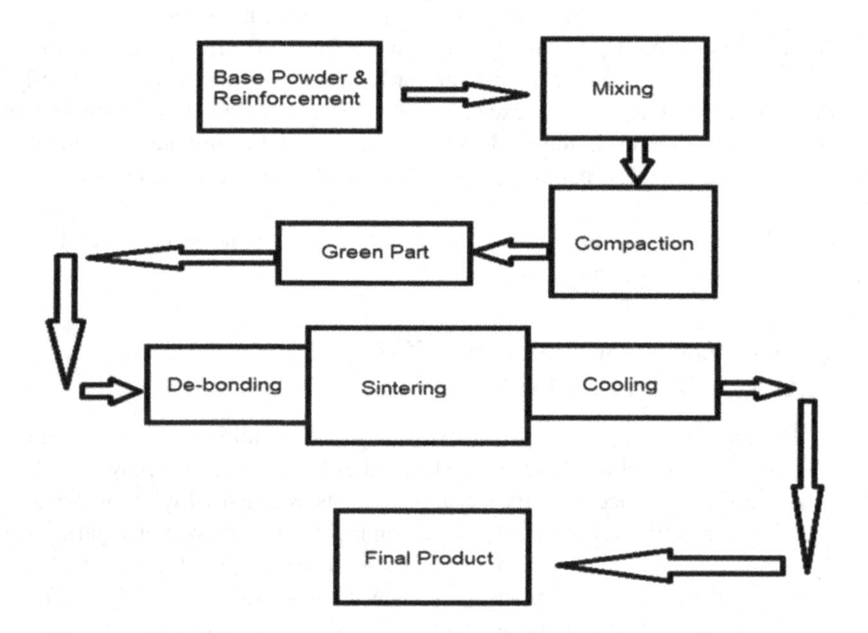

FIGURE 7.1 Conventional PM scheme.

Metal Injection Molding Basics:

- Blend metal powders with polymer binders to create a viscous solution.
- Inject the solution into a molding machine (imagine a toothpaste tube filled with metal powder that's injected into the die cavity).
- Apply heat to remove the binder.
- Sinter at higher temperatures to form metallurgical bonds and densify the powder.

INDICATORS FOR CHOICE BETWEEN CONVENTIONAL PM (POWDER METALLURGY) AND MIM (METAL INJECTION MOLDING)

1) **MIM is excellent for complex shapes.**
   MIM is an economically viable solution for making parts that can benefit from an intricate design for high volume of production. The ability to create an essentially liquid feedstock for injection mold, designing and producing molds with complexity is an essential requirement to adopt MIM route for producing parts in optimum volume. This can be easily compared with the parts produced through the routes of PM using metal dyes by taking metal powders. Very often the MIM route produces parts of same metal with better physical properties than those produced through the PM route.
2) **MIM requires high temperatures to sinter.**
   One of the big downsides to MIM parts is the heat requirements for sinter ing. Heating these parts to sinter them **can add substantial cost** to the manufacturing process.
3) **MIM parts go through approximately 25% shrinkage.**
   A large part of the volume of MIM feedstock is made up of binders – much of which needs to be removed during sintering. That means you lose around 25%.
4) **MIM is expensive and quality control is critical.**
   MIM is 5–10 times costlier than PM. The MIM tools and raw materials are expensive. It demands a tough quality control plan. Debinding and sintering are most crucial steps in MIM.

   The properly controlled MIM process has been proved to be very successful in producing molded parts made of 316 L stainless steel that has resulted in the following mechanical and physical properties:

   - Ultimate tensile strength 75,000 psi
   - Yield strength 25,000 psi
   - Elongation 50%
   - Hardness 65 HRB
   - Density 7.6 g/cm^3

7.1 METAL INJECTION MOLDING (MIM)

The Evolution of Metal Injection Molding (MIM)

- MIM is a very recent technology not older than 40 years. The technology is seen as a combination of plastic injection molding and traditional powder metallurgy.
- Indeed, one similarity between MIM and plastic injection molding is observed when the material is fed into a heated barrel, mixed, and pushed into a mold cavity where it cools and then hardens to the mold die cavity shape.
- As far as traditional powder metallurgy is concerned, the similarity occurs in the way the procedure is implemented. This technique can compact a lubricated powder mix in a rigid die by uniaxial pressure, eject the compact from the die, and sinter it.
- Furthermore, MIM is also seen as a branch of a broader area, powder injection molding (PIM), that involves the use of both metallic and non-metallic powders in the fabrication of small-to-medium-complex-shaped parts in large numbers.

Metal Injection Molding (MIM)

- It is a metallurgical process in which finely powdered metal (<20 μm) is mixed with a binder to create the necessary input material for the component. The feed so created is then injected into a die cavity to create a shaped and solidified component. This process enables high volume, complex parts to be manufactured in a single step.
- From cost and complexity, MIM is considered suitable for the mass production of components used in automotive, electronics, and small arms. As a general rule, a break-even is achieved if the production requirement is 10,000 or above a year.
- It is well suited for components with mass as low as 0.05 grams (0.002 ounces) to 250 grams (8.8 ounces). Wall thickness is typically about 0.5 mm (0.02 in).
- Tolerances are on the order of ±0.3 to 0.5%, albeit specific dimensions can be held as close as ±0.1%.

Injection molding is a manufacturing process for producing parts by injecting molten material into mold(s). Injection molding can be performed with a host of materials, mainly including metals (for which the process is called die-casting), glasses, elastomers, and most commonly thermoplastic and thermosetting polymers (Figure 7.2).

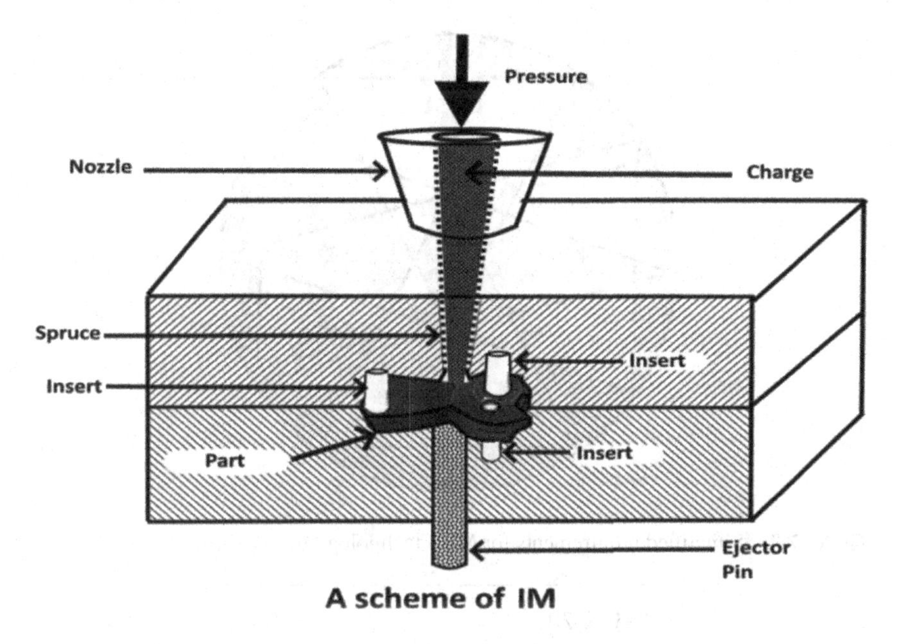

FIGURE 7.2 Injection molding.

Criteria that Decides MIM to be a Preferred Option

- Components with highly complex geometries
- A wide selection of materials to meet specific product requirements (it is even possible to have two different types of material integrated into a single part)
- Need to integrate two or more separate parts into a singular, fully integrated part
- Controlled porosity for lubrication and filtration applications
- Good surface finishing (as-sintered around Ra 0.8 um)
- To minimize or eliminate secondary processes such as machining
- To achieve a low-cost mass production process

Low material wastage – 95–98% of raw material is shipped as a finished product.

Cost
Performance
MIM
Volume
Complexity

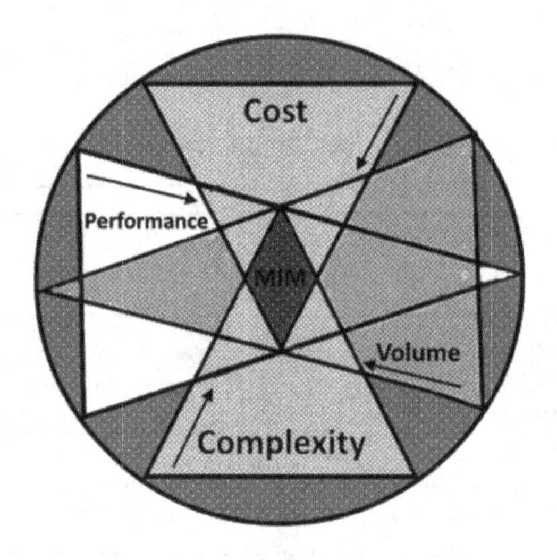

FIGURE 7.3 Simplified requirements for MIM technology in small arms.

TABLE 7.1
Tolerances of Plastics

Molding Type	Typical [mm]	Possible [mm]
Thermoplastic	±0.500	±0.200
Thermoset	±0.500	±0.200

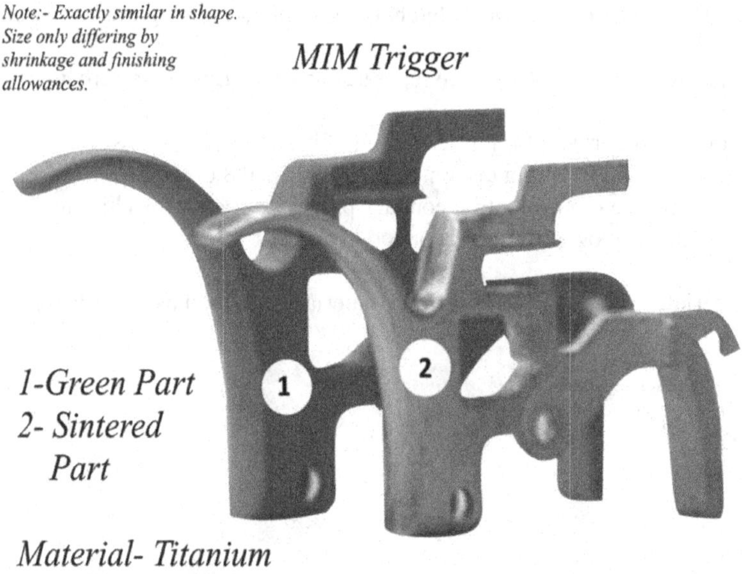

Note:- Exactly similar in shape. Size only differing by shrinkage and finishing allowances.
MIM Trigger
1-Green Part
2- Sintered Part
1
2
Material- Titanium

FIGURE 7.4 Application.

Figure 7.3 indicates that MIM is the best option for adequate performance, complexity, and the requirement of necessary production volume, which must be in demand for producing the parts in optimum cost.

Table 7.1 shows the maximum achievable precessions in producing parts by plastic molding. Comparable/ better precession can be obtained easily for producing the metal parts through the MIM route, which can be subsequently finished through metal finishing operation to more precession parts which are often called for in the design of small arms components.

The part shown in Figure 7.4 is a MIM molded trigger of a firearm. Its complex features are many angular surfaces and several outside radii. A dozen of functional features and surfaces are geometrically controlled in concentricity, profile, and tolerances.

The Capability of the MIM Process

The process can manufacture products from any solid material that ranges from the hardest carbide, pure oxides, natural minerals, and metals. It can also use complex composite synthetic and their various combinations as basic input materials.

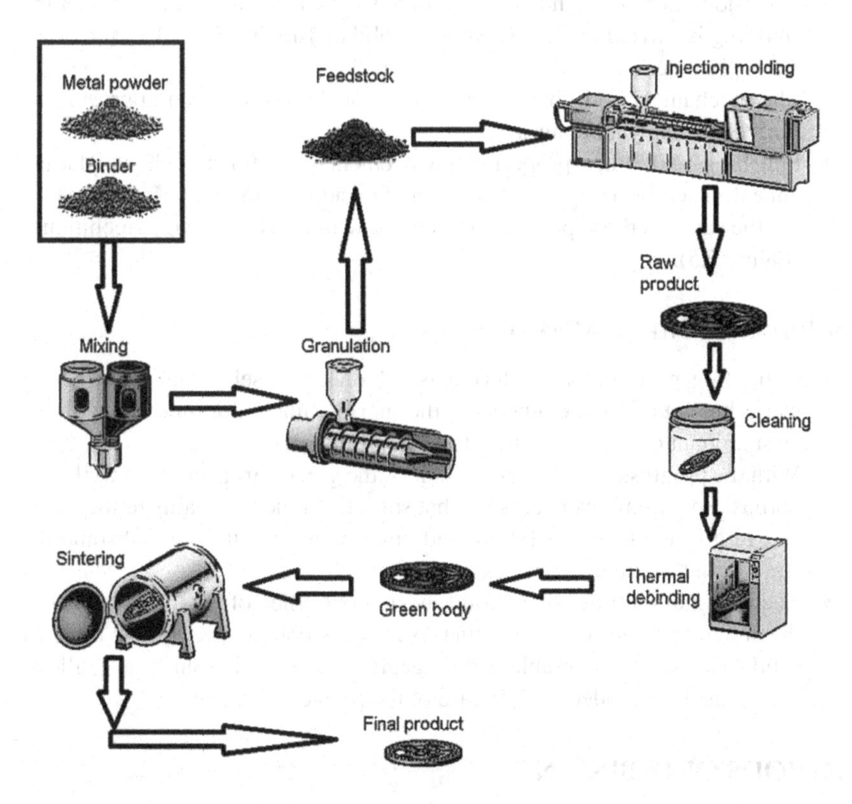

FIGURE 7.5 Metal injection molding (MIM) flow diagram.

Process Steps

- Metal powders are combined with polypropylene and wax binder as the first step to create the feedstock.
- Then the feedstock is injected as a liquid into a mold using plastic injection molding machines.
- The output from the mold is the green part which is cooled and ejected from the mold.
- In the subsequent step, the major portion of the binder is removed from the green using a solvent and a catalytic process in a thermal furnace.
- The debinded and thermally processed part is called the "brown" which is porous (40% volume air) and fragile.
- In sintering, the powder is heated to a temperature near the melting point. The furnace atmosphere is inert and protective. The particles are densified by the action of the capillary forces of the partially melted material.
- The MIM parts are sintered at a high temperature that causes partial melting and holds the metal particles strongly together. A temperature of 1,400°C may be required for materials like stainless steel.
- Diffusion rates are high leading to high shrinkage and densification. If the sintering is carried out in a vacuum at solid density of 95–99% of the original is possible.
- The mechanical and physical properties of the product can approach the limit of the original material.
- Finishing and other property enhancement operations such as plating, annealing, carburizing, Passivation, and nitriding for the MIM components are the same as those parts manufactured with other fabrication techniques (Figure 7.5).

The Important Steps in MIM Debinding

- Debinding parts before sintering is a balance of selectively eliminating some, but not all, of the binders in the shortest amount of time and with the least amount of damage to the structure of the parts.
- With the progressive removal of binders, the green part becomes fragile and porous. So, it remains necessary that sufficient binder remains in the green part to give it adequate strength and dimensional stability to withstand the stresses in the sintering process.
- The minimum left out structural binders are burned off in the high heat of the sintering furnace and the final product is obtained into their finished solid state with considerable shrinkage. In the part design shrinkage, allowance must be provided to take care of the problem.

METHODS OF DEBINDING

- There are several methods for debinding green parts.
- Thermal debinding – A heating process to remove the wax binders and any additives. (A long process that takes 24 hours).

- Catalytic debinding – nitric or oxalic acid is used as a catalyst to break down the binders. (A quick process can be completed in few hours).
- Debinding by vapor degreasing.

APPLICATIONS OF MIM

- The product particulars such as part numbers manufacturing date can be incorporated into the design. Complex 3D geometries can be produced by reducing material waste and cost. Ninety-five to ninety-eight percent of original material density can be imparted in the final product.
- The ability to combine several operations into one process ensures MIM is successful in minimizing lead times as well as costs when production volume is high.
- The MIM process may be considered as an eco-friendly technology due to the significant minimization in wastage compared to conventional manufacturing methods such as 5 axis CNC machining.

QUALITY CONTROL ISSUES IN OPTIMIZING DESIGN FOR MIM PRODUCTION

Desirable Features

- Gradual section thickness changes
- Range of component thickness, 0.1–10 mm
- Low mass
- One flat surface for support during sintering
- Draft – where and when
- Corner breaks and fillets
- Holes and slots
- Undercuts – external and internal
- Threads
- Ribs and webs
- Knurling, lettering, and logos (Figure 7.5)

Allowed Design Features

- Axisymmetric, asymmetric, prismatic, square, and freeform features
- Cantilever and asymmetric shapes
- Protrusions, bosses, and studs
- All shapes of holes are allowed, including holes at angles to one another
- Undercuts, grooves, slots, depressions
- External teeth, external or internal threads

Features to Avoid

- Inside closed cavities
- Undercuts on internal bores

- Very sharp corners or edges – desired radius greater than 0.05 mm
- Long pieces without a draft or taper to allow ejection
- Holes under 0.1 mm in diameter
- Walls thinner than 0.1 mm

Size Limitations for MIM

- Since molding, debinding, and sintering are easier and faster with thin-walled components, the overall size is not much constraint if the thickness of the design component is appropriate. The ultimate strength and quality of the component is dependent on the effectiveness of debinding and sintering.
- MIM parts of mass (12–17 kg), length (up to 300 mm), and section thickness of (125 mm) have been successfully produced.
- From a cost standpoint, the general driver is toward nominal 20-micrometer powders, negating the production of sharp edges, despite the tooling.
- The limitations on the corner and edge radii are 0.05 mm and 0.075 mm respectively– are even more realistic. Sharp edges are usually formed by grinding after sintering.

The Largest MIM Component

The quick answer is, in most cases, less than 160 grams. But this is not correct as many geometric attribute matters.

Mold Restrictions on Part Size

The size of the part is restricted by the capacity of the mold and the capacity of the plant equipment. The process itself is not a limitation.

Part Complexity

The complexity of a component helps to dictate which process should be utilized for mass production to be cost-effective and efficient.

The Efficiency with MIM Materials

- MIM feedstock – which can be custom – is inherently more expensive than your average recycled aluminum, so it is not surprising that material cost does play a role in the MIM decision-making process.
- Design the components to optimize the weight of the component and only use as little material as possible to create the component.
- With the MIM process, complexity can be added to the component without having to add material. While a larger part that is machined often results in a lot of scrap and waste.

Part Size Related to Sintering and Debinding

- Next to molding, sintering and debinding furnaces have strict guidelines regarding mass loading for each batch and size of components so that the binding material is removed at a proper and precise rate.
- The larger and thicker the parts, the fewer components you can put into the furnace at one time, and the longer it takes to sinter and debind.
- Remembering that time is money, shorter cycle times are far more cost-effective; therefore, less mass (part volume) usually equates to less process cost/time.
- In small arms, MIM offers a low-cost solution to small complex components that would otherwise have to process through expensive secondary operations.
- MIM process can utilize a broad range of materials. Conventional processes incur considerable material waste and are normally restricted to mono-material construction. MIM is an efficient option for the fabrication of components made of expensive or special alloys. Sometimes MIM is a viable option for a very thin wall specification of 100–200-micron wall thickness.

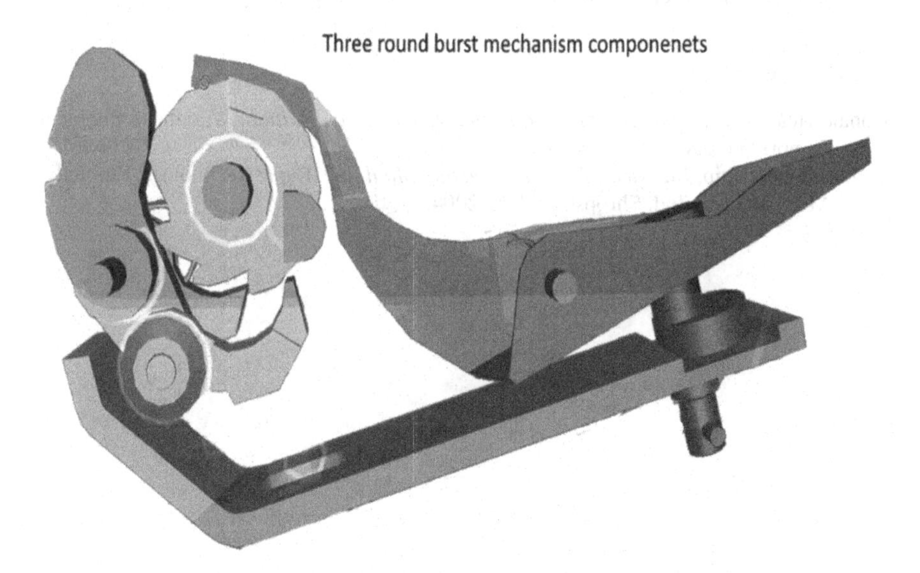

FIGURE 7.6 Example of parts that can be manufactured using MIM in Firearms.

- Part designs can be limited when constrained by traditional metal-working processes. However, with MIM design engineers having the freedom to create parts by placing material only where it is needed for function and strength, the final result is a complex shape that uses less material and does not have to be machined.

- Figure 7.6 shows the parts of a trigger mechanism of an assault rifle. Almost all the parts of this sub-assembly meet the criteria for the choice of MIM route for their production in medium-to-high volume for economic and quick production with desired precession.

QUALITY CONTROL IN THE MIM PROCESS AND MANUFACTURING

The final product quality in the product manufactured through the MIM process is dependent on the process parameters and materials employed in manufacturing. Hence, necessary care should be taken in respect of the following:

- Gating – types and location
- Sink and knit lines
- Flash and witness lines
- Interchangeable mold inserts
- Choice of binder and solvent for debinding
- Control of sintering temperature and environment
- Control of "green and brown" density to control porosity

BIBLIOGRAPHY

Donald Heaney, *Handbook of Metal Injection Molding- 1st Edition*, Woodhead Publishing, Cambridge, UK, 2012.

Vannessa Goodship, *Practical Guide to Injection Moulding*, Rapra Technology Limited and ARBURG Limited, Shropshire, UK, 2004.

8 Surface Treatment of Small Arms

SURFACE TREATMENT OBJECTIVES

- Corrosion protection
- Durability and longevity
- Wear resistance
- Enhance fatigue strength
- Minimize friction
- Improve aesthetics

AVAILABLE OPTIONS

- Gun bluing (**Black Oxide**)
- Gun anodizing
- Cerakote coatings for guns
- Ferritic nitro carburizing firearms
- Quench polish quench (QPQ)
- Chrome and nickel plating
- CVD and PVD coating

PRIMARY STEPS TO SURFACE FINISH

- Sandblasting
- Roto finishing
- Barrel finishing
- Parts cleaning
- Degreasing
- Passivation
- Lapping
- Build-up welding

Gun Bluing

- Bluing of steel is a traditional way to protect firearms from corrosion while at the same time reducing glare.
- It is achieved by an electrochemical reaction that changes iron to black oxide.

DOI: 10.1201/9781003199397-8

- This transformation gives treated steel some corrosion protection but requires frequent oiling to keep rust at bay.
- On the plus side, this type of treatment does not change the dimension of the piece, so there's no need for gun parts to accommodate that increase.
- There are several types of bluing methods used for different applications.
- Cold bluing changes the color of steel to an attractive dark gray and is best used for small touch-ups or worn areas on a gun. Selenium oxide-based compound is used at room temperature.
- Hot bluing is used most often in the manufacturer's setting, resulting in a stunning blue-gray finish. Alkali salt was used at 275–300°F.
- Rust or fume bluing. This method is known as the Taj Mahal of this genre of gun finishes, as it offers unparalleled corrosion resistance by initiating a reaction that converts rust-loving metals magnetite, thus preventing the beginning of rust itself.

Parkerizing

- An alternative to bluing, this chemical phosphate conversion coating results in an anti-reflective gray to black finish with very little change in dimensions.
- It is slightly more effective at corrosion resistance than bluing but requires frequent oiling to retain this attribute.
- Often coated with epoxy or molybdenum finish to make the gun parts self-lubricated.

Cerakote, Duracoat, and KG Gunkote

FIGURE 8.1 M4 Carbine – gun coating.

Spray-on Method

- Guns are coated with spray-on gun-coating techniques that are branded as Cerakote, DuraCoat, and KG Gunkote. An example of gun coating is shown in Figure 8.1.

- Each differs in composition, and all processes cause a slight increase in dimensional thickness.
- Duracoat is a coating technology specially developed for firearms. It is a two-stage technology and can withstand 300 hours of salt-spray exposure.
- KG Gunkote was developed for rigors of military use and is similar to other spray-on finishes in its capacity to protect against chemicals.
- Its corrosion resistance was proven by passing 500-hour salt exposure tests, so it's a finish that handles just about anything.
- Cerakote is a polymer–ceramic composite that can be customized in a myriad of colors, offering excellent corrosion protection. In salt-solution testing, it exceeded the 550-hour mark, which is well over the minimal military standards.
- It can be applied to many different surfaces and is one of the toughest, most durable finishes marketed today (Figure 8.1).

Many firearm dealers are so confident in the longevity of this finish that they offer lifetime workmanship warranties against peeling or cracking.

Nickel Boron

Electroless Deposition

- This chemical coating requires no need for electrochemical reactions, leading to a more uniform finish.
- It's low-friction and diffuses heat well. It is perfect for intricate parts found in internal firearm components such as bolt-carrier groups.
    - The solution uses sodium borohydride or alkylamine borate with a source of nickel ion and a control-chemical which is unique for each kind of metal deposition.
    - It requires dimensional compensation when being used on moving parts.

Nitride Coating (QPQ)

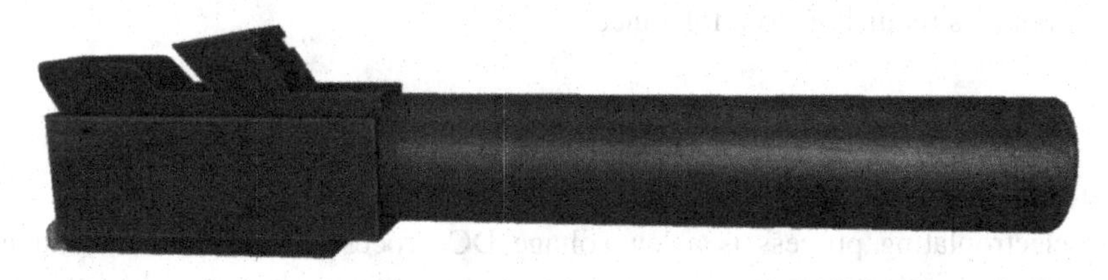

FIGURE 8.2 Nitride coating of a pistol barrel.

- Ferritic Nitro carburizing – a modified form of nitriding not carburizing. An example of Nitride Coating in pistol barrel is shown in Figure 8.2.

- The process introduces nitrogen and carbon into steel or any iron-containing alloys.
- The more effective versions of it called Quench Polish Quench (Melonite, Tufftride, or Tenifer) are surface transformation treatments that result in no dimensional changes, so they are well suited for internal components.
- QPQ is carried out in the salt bath layer of iron and iron-nitride is formed. It is polished and again immersed in a salt bath and a final layer of magnetite is formed. Alkali cyanates are used in the salt bath. Treated surfaces are hard and very wear-resistant with a decent amount of lubricity.
- A hardness of 64 HRC is achieved against 70 HRC of diamonds (Figure 8.2).

PVD or CVD Coating

- PVD and CVD offer a tough, micro-thin surface with great lubricity.
- Colors can be customized to a certain extent, so it is a good choice for those who want a thin, durable finish with the color of choice.
- It is a high-temperature process (600–800°F), therefore has to be selectively used on intricate parts that are liable to warping at high temperature.

Anodizing

- This finish is second only to diamonds in hardness.
- An anodized finish is created by an electrochemical process similar to bluing, but results differ in one specific way.
- Dimensional increase. Treated parts increase in size, so adjustments must be made to accommodate that result.
- A wonderful quality of anodized metal is that its porosity lends to better paint or glue adhesion, so much so that some guns are anodized before a spray-on color coat is applied.
- Chromic acid-based anodizing at room temperature – **Bengaugh Stuart** process.
- Sulfuric acid-based anodizing at 32° F – **Hards** process.
- Aluminum containing a high amount of Mg and Cu is unsuitable for components requiring close tolerance.

Electroless vs Electrolytic Plating

The Electro Plating Process

The electroplating process is a low-voltage DC process in which the metal ion assumes a positive charge and deposits on a substrate that works as a cathode to form a metallic layer.

Electroless Plating Process

The electroless plating process is much simpler as it uses no electricity and requires no extra equipment. In the electroless plating process, the part or substrate receives its deposit via an autocatalytic reaction via a reducing agent. The reducing agent interacts with the metal ions making the deposition possible. The process is completed with an application of anti-oxidation chemicals which improves the part's resistance to corrosion and friction, prolonging its life.

The advantages of electroless nickel over electrolytic nickel plating:

1. Uniform deposit thickness
2. Superior corrosion resistance
3. Variable magnetic properties
4. Harder as plated deposits and hardness achievable is 90% of hard chromium
5. Improved deposit tribology

Electroless Nickel Plating

Electroless plating is the selective reduction of metal ions only at the surface of the catalytic substrate that is immersed into an aqueous solution of desired metal ions. A continuous deposition on the substrate through the catalytic action of the deposit itself takes place. Electroless nickel (EN) plating is the most important catalytic plating process. The main reasons for its commercial and industrial use are the unique properties of the EN deposits. The properties of EN deposits depend on the formulation and operating conditions of the EN plating bath. EN plating can be applied to both conductive and non-conductive surfaces.

Typically, the constituents of an EN solution are:

- A source of nickel ions
- A reducing agent
- Suitable complexing agents
- Stabilizers/inhibitors
- Energy

BIBLIOGRAPHY

D. R. Gabe, *Principles of Metal Surface Treatment and Protection*, Pergamon Press Limited, Oxford, England, 1978.

P. Hodgeson (Eds.), *Quenching and Carburising – Proceedings of the Third International Seminar of the International Federation for Heat Treatment and Surface Engineering*, The Institute of Materials, London, UK, 1993.

Rory Wolf, *Plastic Surface Modification. Surface Treatment and Adhesion*, Hanser Publications, Cicinnati, Ohio, USA, 2010.

9 Common Defects

COMMON GUN MALFUNCTIONS

Gun malfunction is the condition where it is not possible to fire a gun smoothly and safely. And the weapon goes through unintentional stoppages and fails to fire. A gun generally malfunctions on account of the followings:

1) Failure to fire
2) Failure to feed in single fire mode
3) Failure to extract
4) Failure to eject
5) Rim lock
6) Hammer follow
7) Slam fire
8) Stovepipe or failure to feed consecutive rounds continuously in course of auto mode
9) Double feed
10) Out of battery
11) Recoil spring defect
12) Spent case and firing pin defect

Failure to Fire

- Those shooting rimfire rounds may come across this every once in a while, as that type of primer is slightly less reliable than the traditional center-fire primer. This happens in a center-fire cartridge due to either a broken firing pin or due to the inadequate impact of the firing pin.
- If misfire happens: First, keep the gun pointed down range for at least a minute. After a minute has passed there is a small risk of the defective cartridge accidentally going off. After this, remove the round from the chamber and unload the gun.
- If the primer has a well-defined indentation on it – this indicates a faulty round. If there is a shallow mark on the primer, the gun is probably dirty and the firing pin is suffering from carbon buildup/broken.

DOI: 10.1201/9781003199397-9

Failure to Feed

- Failure to feed occurs when the cartridge does not fully sit within a chamber. Semi-automatic firearms can experience a failure to feed issue due to the following:
- Damaged magazines
- Weak magazine springs
- Dirty or greasy chamber
- Improperly seated magazines
- Damaged or faulty cartridges

Failure to Extract

- Failure to extract is the inability of the firearm to remove the spent cartridge from the chamber, followed by double feeding of a new round from the magazine. This is a serious malfunction that can cause an unsafe situation.
- To clear the malfunction, the magazine from the gun should be removed; after that point the gun in a safe direction and rack the slide two or three times to completely clear the chamber. After completing a visual inspection of the chamber and magazine, ensure that the gun is empty.
- A failure to extract is also caused by damaged or low-quality magazines or weak magazine springs. This is also caused by shooters who do not have a firm grip on the gun or who ride the slide as it moves forward. "Limp wristing" frequently causes failure to extract. Such types of repeated failures indicates poor design and mechanical impedance mismatch.

Failure to Eject

- A dirty gun or a corroded chamber can cause failures to eject, similar to the lack of a firm grip by the shooter. If this situation occurs, one should do what is called the "Tap Rack Bang" drill.
- In the process of the "Tap Rack Bang" drill, tap the magazine to ensure that it is perfectly seated. Then rotate the firearm slightly so that the ejection port is at an angle to the ground, then rack the slide firmly. This will eject the spent cartridge and allow a new cartridge to enter the chamber. The "bang" means pulling the trigger when the gun is safely aimed downrange.
- This malfunction is also known as a "stovepipe". This occurs when the spent cartridge is trapped in the ejection port.
- In assault rifles with the provision of side ejection of spent cases, the requirement of forward ejection is a primary functional requirement for a weapon. Therefore, the proper design of the extractor spring and extractor is necessary to ensure this function.

Rim Lock and Broken Extractor

- When the extractor fails to get over the rim of the bullet to catch it over its neck and the gun fails to be in the battery, the problem is called a rim lock. It is a common problem for bullets with large rims such as 7.62 × 54 mmR.
- The broken extractor may be a frequent case in automatic weapons while using steel cartridge cases.
- Make sure to keep your chamber clean frequently to remove the deposits of leaked gases.

Hammer Follow

- Hammer follow occurs when the disconnector allows the hammer to follow the bolt and firing pin into the battery, sometimes causing the firing mechanism to function without pulling the trigger.
- This is usually a result of extreme wear or outright breakage of firing mechanism components and can result in an uncontrollable "full-auto" operation, in which multiple rounds are discharged following a single pull of the trigger.

Slam Fire

- A slam fire is a premature, unintended discharge of a firearm that occurs as a round is being loaded into the chamber.
- If the firing pin fails to retract into the bolt as the bolt moves forward in the loading part of the cycle, the protruded firing pin causes unintended fire and this is known as slam fire. This may also happen due to the momentum of the free-floating firing pin during the loading of the gun.
- Similar to a hammer follow malfunction, this can result in an uncontrollable "full-auto" operation.

Stovepipe

- A stovepipe is a defect when an empty cartridge case fails to eject out and get caught in the ejection port. This is a defect common to pump-action, semi-automatic, and fully automatic firearms that fire from a closed bolt.
- Defective extractor or ejector or improper hold of the shooter are common reasons for stovepipes to occur.
- When cartridges are not sufficiently powerful to fully cycle the action, or there is an inherent mismatch in mechanical impedance between static and moving parts due to poor design stovepipe occurs.
- This also results after some prolonged firing, when the supply of gas for automatic action decreases due to the fouling of the gas port which results in a decrease in relative velocity between block breech with career assembly

and the frame. The block breech fails to cross over the magazine fully to take up a new round.
- This is one of the most serious defects when combating in close quarter battle (CQB) role where a malfunction means life and death.

Double Feed

- When two rounds are picked up from the magazine, one by the bolt and another by the bottom of the carrier and attempt to feed into the chamber at the same time, a problem of double feed is said to have occurred. This may be due to a bad magazine as well.
- This may be also due to the poor quality of the bolt carrier in terms of its quality of contact surface and defect in the leading-edge chamfer which may act as a feeding horn.

Out-of-battery

- When the slide/bolt is in the normal firing position, the firearm is said to be "in-battery". It is "out-of-battery" if the slide/bolt/action is not perfectly seated in the normal firing position.
- The firearms are provided with safety features so that they are not capable of firing when out-of-battery. Out-of-battery leads to a dangerous situation.
- The cartridge casing itself cannot withstand the pressure of firing. It is dependent on the walls of the chamber and the bolt face to hold the pressure.
- In any of the conditions of the round not fully chambered or the firearm is out-of-battery, or the bolt face is not supporting firmly the rear of the cartridge, and if the round is fired in this situation, the case will burst, causing high-pressure hot gasses, fragments of the casing and bits of the burning powder itself to be thrown at high speed from the firearm into the receiver.
- This can be a serious hazard to the shooter and any bystanders.

Recoil Spring Defect

- A weak recoil spring will not store sufficient energy to cycle properly and will cause a failure to feed problem. The weak spring will also allow the significant impact to be transmitted to the rear of the receiver and cause damage to the receiver.
- Recoil spring assembly has a life of a specified number of rounds. And should be replaced on reaching the round count.

Spent Case and Firing Pin Defect

Be observant of the spent casings. There could be an issue with the gun's firing pin. This can be easily seen by looking at the primer area of the casing. If it hit anywhere other than the center (or rim for rimfire guns) or didn't strike firmly enough, you could have a bent or misaligned firing pin.

BARREL DEFECT

Identifying Cracks in Gun Barrels

- Gun is subjected to repeated impact on firing, which induces fatigue stress. The prolonged exposure to fatigue stress weakens the barrel material and that may result in a catastrophic barrel burst.
- This is typically prevented by assigning a maximum round count to the barrel set (say 10,000 rounds for intermittent firing).
- This is a mandatory replacement point.

Barrel Erosion

- Barrel wear in a center-fire rifle is almost exclusively due to throat erosion (cracks and roughness in the first 3–4 inches ahead of the chamber caused by heat, flame, and pressure).
- The erosion of gun barrels leads to reduced gun performance and availability and increases the expense of barrel replacement over the lifetime of a gun system.
- Wear in the gun barrel usually increases the bore diameter near the commencement of rifling.

CARTRIDGE MALFUNCTIONS

- **Case head separation –** When there is excessive friction between the cartridge case and the chamber wall, hard extraction occurs and if the case is not strong enough, the head comes off the case and the body remains inside the chamber. This can also happen with worn-out chambers in which the case expands beyond the elastic limit and fails to spring back into shape on the exit of the propellant gas.
- **Dud –** A dud is said to have occurred when a cartridge misfires. It should be handled carefully because a hang fire can happen.
- **Hang fire –** A delay in firing upon the impact on primer is called a hang fire.
- If hang fire occurs, keep the firearm pointed downrange or in a safe direction for a minute, then remove and safely discard the round, because the functioning of a round outside of the firearm, or in the firearm with the action open, could cause a serious explosion hazard.
- **Squib load –** This is an extremely dangerous malfunction that happens when the projectile does not have sufficient energy to leave the barrel. This may be due to oversized projectile or insufficient propellant. In semi-automatic

and automatic weapons, the subsequent round impact on the squib load which can cause a catastrophic failure of barrel burst. A squib load should never be cleared by firing either a live round or a blank round.

PREVENTION OF MALFUNCTIONS AND CATASTROPHIC FAILURES

- Malfunctions occur due to poor design and cannot be easily be avoided.
- Many malfunctions can be prevented by proper cleaning and maintenance of the firearm.
- Proper selection of material, processes, and choice of quality in design can improve the reliability and decrease failure probabilities of the weapon.
- The sustained firing in the weapon produces a lot of heat and stress which deforms and damages the weapon very rapidly, so the limit of the maximum number of rounds in sustained firing must be specified for enhanced life.
- The barrel is the most heated and stressed part, so the material of the barrel should have high torsional and flexural rigidity for avoiding catastrophic failures such as barrel burst or bending and twisting.
- The processes like quench polish and quench (QPQ) should be used for trigger mechanism parts wherever possible for trouble-free operation.

10 Catastrophic Failure

PRIMARY REASONS FOR CATASTROPHIC FAILURES

The weapon system behaves as an intermittently fired vented pressure vessel. It delivers a mechanical output in the form of projectile energy. The pressure development to a great extent depends on the driven load. The driven load is determined by the inertia of the bullet and the force required to extrude the bullet through the barrel bore.

In addition to normal operation, two extreme conditions/events are likely to happen within the barrel:

- A blank firing condition where there will be a practically minimal driven load which will result in the development of very low pressure in the barrel which is not sufficient for the generation of cyclic operation in using any of the operating principles.
- A perfect bullet lodge may occur just even at the neck of the chamber, resulting in a closed pressure vessel and consequently a constant volume combustion. This is the most serious situation as nearly 90% of the propellant energy will get converted into pressure energy. It practically gives rise to peak pressure which may be many times more than the allowable design pressure. This situation leads to the catastrophic failure of a barrel burst.
- Due to constant use, copper or lead buildup also takes place inside the barrel bore and creates gradual local constrictions in the bore. This phenomenon is predominant near the gas port. The extra resistance created due to constriction gives rise to pressure buildup which makes it the safe design pressure and barrel bulging cases may come up.
- The peak pressure is dependent upon the projectile section density. Over grain bullet used during manual loading will also increase the barrel pressure. Since the acceleration of the projectile is of the order of tens of thousands of g, the increase in pressure due to an increase in bullet grain will also create dangerous pressure leading to barrel failure.
- The barrel is subjected to repeated shock load in the course of its life and therefore subjected to fatigue stress. Though the material of the barrel is carefully selected, the repeated fatigue stress induces brittle nature in the barrel material. So, each barrel has to be used only up to a specified number of firing rounds for safety reasons.

Frequently, it is noted that in some ill-conceived designs there is an inadequate arrangement of designing the propellant gas flow-path. This, coupled with continuous fouling of the firearms in the course of firing, may further reduce the venting arrangement of the propellant gases. This results in an abnormal buildup of gas

DOI: 10.1201/9781003199397-10

pressure in the enclosed part of the body housing which results in the bursting of the weakest component, namely the magazine, or in extreme cases the collapse of receiver housing mostly made of polymer or weaker materials.

UNCONTROLLED BURNING OF THE PROPELLANT

The ideal propellant on ignition must deflagrate and should not detonate. The improperly filled cartridges may fail to burn in a controlled manner and result in detonation, and the faulty ammunition can thus become a reason for catastrophic failure. Also, abnormal variation in the charge mass and density or increase in the bullet mass can cause the pressure to increase abnormally, which can lead to dangerous catastrophic failure. This aspect should be specially taken care of by those weapon users who try to load the weapon with their own customized ammunitions.

Cartridge head space (CHS) – It is one of the important parameters for reliable firearm operation. The correct CHS means holding the cartridge firmly supported against the chamber shoulder. It has to prevent even minute slippages of the cartridge case from within the chamber. If the slippage of the cartridge happens, the weaker part of the cartridge case will become unsupported at the moment of firing and this will result in a split head cartridge case, which will damage the weapon and render the weapon unserviceable till the spent case is taken out of the chamber and the CHS is finally readjusted.

The resultant explosion may also invade the inside of the housing mechanism resulting in either the burst of the magazine or damage of the housing mechanism.

Another reason for catastrophic failure is the use of manually imprudent loading of the cartridge, which is often carried out by enthusiastic users. The internal ballistics requires very precise loading of the case with the propellant. Even a minimal variation beyond the permissible limit can give rise to either a situation of overpressure due to excessive loading or a condition of detonation due to underloading. Both produce damaging stresses to the weapon (Figures 10.1–10.8).

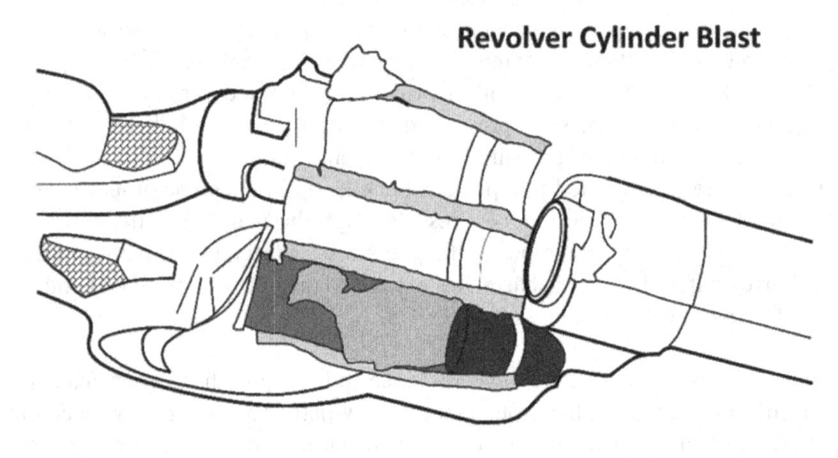

FIGURE 10.1 Revolver- cylinder burst.

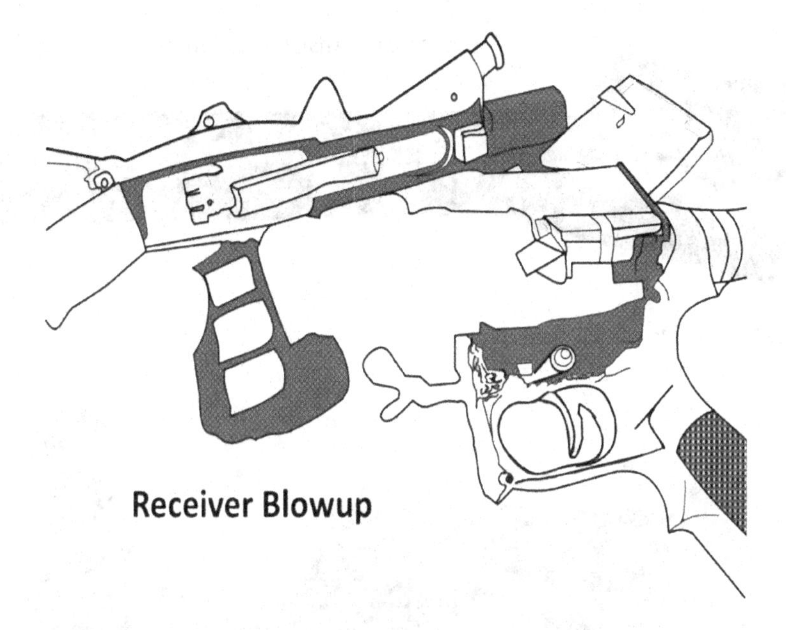

FIGURE 10.2 Lower receiver exploded.

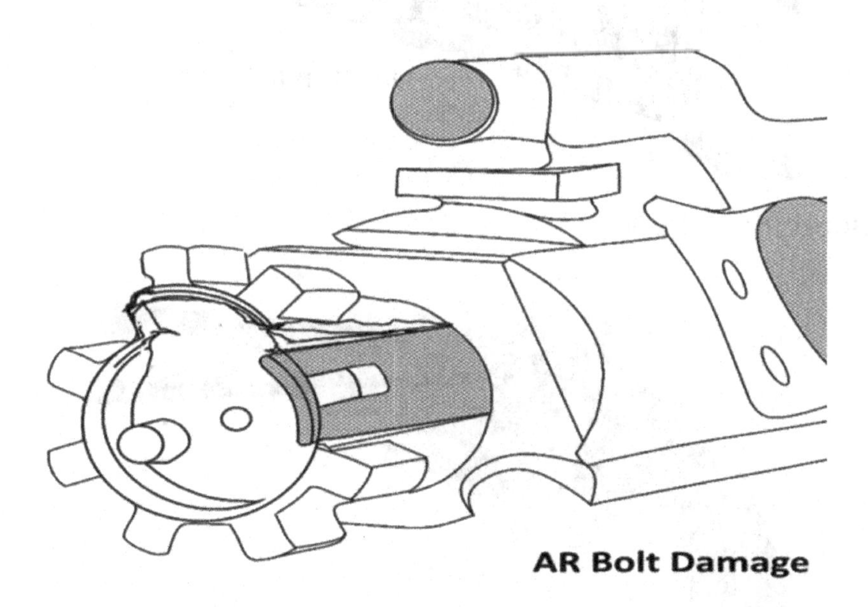

FIGURE 10.3 Bolt damage.

FIGURE 10.4 Pistol chamber rupture.

FIGURE 10.5 Barrel burst.

FIGURE 10.6 Revolver squib load.

FIGURE 10.7 Common stove pipe.

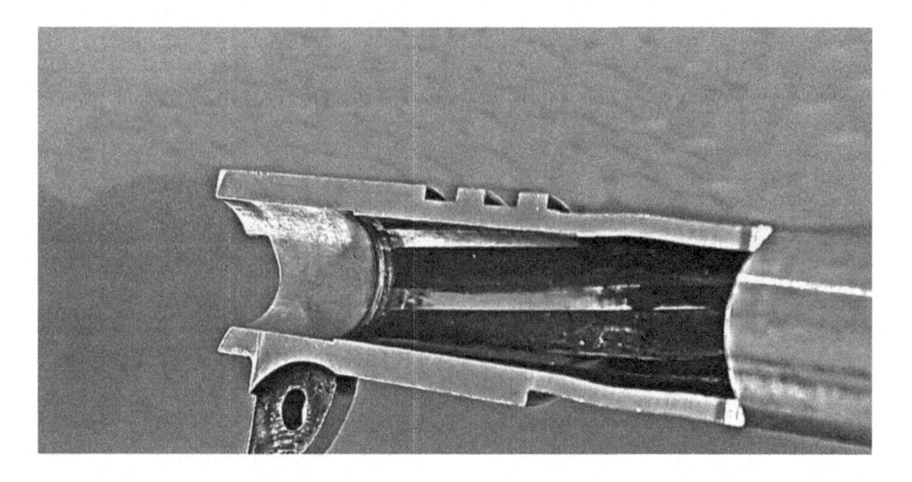

FIGURE 10.8 Barrel bulge.

11 Proof Parameters

GENERAL OBJECTIVES OF FULL AMMUNITION AND FIREARMS TESTING

- Ballistics Testing: Pressure
- Ballistics Testing: Velocity
- Ballistic Testing: Accuracy
- Ballistics Testing: Dispersion
- Strength and endurance
- Safety & Performance

PROOF REQUIREMENT AND TEST PARAMETERS FOR VARIOUS SMALL ARMS

Proofing of a firearm is needed to check the strength of the firearm for safety purposes while it is being used in actual operation. The various parameters tested while doing the proof test are strength, endurance, safety, and performance.

Proof Test

- The Proof test is the test of the strength of a structure by applying pre-estimated stress to validate the design and ensure that it will be safe in service for a specified lifespan.
- In the proof test, the structure is subjected to loads much above that expected in actual service to demonstrate safety and confirm the design margin.
- The design margins and test levels are chosen in such a manner that the proof test is nondestructive. And the structure tested must remain serviceable after the completion of the test. It is generally given a proof mark for certification of the test.

Objectives of Ammunition Testing

- Pressure and velocity measurement
- Proof testing standards (C.I.P. /SAAMI)
- Cook-off testing
- Less-than-lethal testing
- Projectile characterization
- Quality control inspection
- Performance evaluations for accuracy and dispersion
- Corrosive primer testing
- Primer sensitivity testing
- Environmental testing

DOI: 10.1201/9781003199397-11

Small Arms Ammunition Pressure Testing - Civilian Test Methodologies

C.I.P. Method – Commission Internationale Permanente Pour L'épreuve des Armes à feu Portatives, France

- In C.I.P. a drilled case is used to expose the pressure transducer directly to propellant gases.
- The piezo measuring device (transducer) is positioned at a distance of 25 mm (0.98 in) from the breech face when the length of the cartridge case permits that, including limits.
- For too short a cartridge case length, pressure measurement is taken at a chamber-specific defined shorter distance from the breech face that depends on the dimensions of the case.

SAAMI Method – Sporting Arms and Ammunition Manufacturers' Institute, Connecticut, USA

- In the SAAMI pressure testing protocol, a conformal Piezoelectric Quartz Transducer is used for pressure testing of centerfire revolver, pistol and rimfire cartridges, and centerfire rifle. Also, it uses test barrels that have a hole located in the chamber at a location specific to the cartridge.
- As the cartridge is fired, the cartridge case expands due to gas pressure and presses the chamber walls. The portion of the cartridge case in contact with the face of the conformal transducer exerts pressure on the transducer, which in turn generates a minute electronic impulse that is amplified and results in reading in pounds per square inch (psi).

PROOF TEST DIFFERENCES

TABLE 11.1
Proof Test Differences

SAAMI	C.I.P.
For bottleneck case: Transducer location – 0.175 inches behind the shoulder of the case of a large diameter of 0.250 inches and 0.150 inches behind for small diameter transducers. For straight case: The transducer is located one-half of the transducer diameter plus 0.005 inches behind the base of the seated bullet. If the case diameter at the point of measurement is less than 0.35 inches, then small transducers are used.	A drilled case is used and the transducers are positioned at a distance of 25 mm from the breech face when the length of the cartridge is sufficient. When the length of the case is short, the pressure measurement takes place at a cartridge-specific defined shorter distance from the breech face that depends on the dimension of the case.

The difference in the location of pressure measurement gives different results in SAAMI and C.I.P. standards.

Instrumentation Used in Proofing

- In carrying out a charge development, or Strength of Design (SOD) test, the mass of the propellant and the service pressure is gradually incremented to the required proofing pressure of the weapon system.
- Measurement of chamber pressure is done by the copper crusher or piezo-electric gauges, and the velocity is measured by Doppler radar or photocell counter chronographs.
- The strain and temperature readings are also recorded.
- If required, high-speed photography (synchro-ballistic photography, high-speed digital stills, head on cine, or flight follower) may also be used.

Small Arms Ammunition Pressure Testing – Military Test Methodologies

NATO EPVAT stands for Electronic Pressure Velocity and Action Time.

- NATO certifies 7.62 mm, 5.56 mm, 12.7 mm, and 9 mm using the NATO EPVAT test methods, where pressure testing is included.
- NATO EPVAT testing procedures for the "NATO rifle chamberings" require the pressure sensor or transducer to be mounted in front of the case mouth.
- It is not necessary to drill the cartridge case to mount the transducer. The process is faster.

Firearms Proof Testing

- A firearm's chamber and barrel become a vented pressure vessel for short periods. A proof test is a test where a deliberately over-pressured round is fired from a firearm to verify that the firearm is sound and will not burst on firing. The firearm is inspected after the test, and if it is found to be in perfect condition, then it is given a "proof mark" to certify that it has been proofed.
- A "proof round" is a special assembly of propellant bullet and case and primer designed to be used in proof testing. The "proof shot" is a special bullet designed for a proof round.
- Small arms proof rounds look like normal cartridges, with special identification for distinction. In many regions sale is not permitted without a proof mark.

Proof Loads

- SAAMI recommendation (rifle) – 33–44% overpressure on the nominal rating. It requires single proof firing.

- SAAMI recommendation (handgun) – It uses the same overload as the rifle.
- C.I.P. requirement (rifle) – 25% over the normal rating. C.I.P. requires two firings.
- C.I.P. requirements (handguns) – It uses 30% overpressure over the normal rating.
- Under the British base crusher standards, proof loads ran 30–45% above normal.

OBJECTIVES OF FIREARMS TESTING (ACCEPTANCE QUALITY CONTROL)

Each firearm is individually tested to certify expected performance during their service life following the manufacturers' standard manuals and the parameters measured and recorded for quality certification as under:

- Accuracy and dispersion
- Rate-of-fire
- Barrel erosion and barrel life
- Weapon endurance
- Trigger pull
- Recoil
- Sound overpressure (dB)

The peak pressure and maximum average pressure developed in the firearms are dependent on both variables and attributes of the cartridge and the barrel. The variables of the cartridge that control the pressure include the case volume of the powder chamber, the projectile mass, shape, and size, charge mass, and density. The variables of the barrel that control the pressure are chamber volume, length of the free bore, overall barrel length, number of grooves, groove depth and width, twist rate, and convergence/divergence of the basic barrel bore. Therefore, when we refer to maximum peak pressure, that is with respect to some standard cartridges and standard test barrels. The approximate peak pressure for a few firearms, with dimensions and features of standard cartridge and barrel, are indicated in Tables 11.2–11.4, for the reader's appreciation of the firearm hazards and design issues. For sporting arms, the reader may refer to SAAMI/C.I. P for necessary guidance for design, production, and testing. Table 11.1 explains the basic difference between SAAMI and C.I.P test procedures.

TABLE 11.2
Approximate Average Maximum Rifle Cartridge Pressure

Calibers	Pressure (psi)
7.62 × 39 mm (M43)	44,000
7.62 × 51 mm NATO	59,000
50 BMG	56,000
5.56 × 45 mm NATO	54,000
270 Winchester	66,000
.300 Weatherby Magnum	64,000
.300 Winchester Magnum	63,000
22 Long Rifle (rimfire)	23,000
.22 Short (rimfire)	22,000
.22 WRF	18,000

TABLE 11.3
Approximate Average Maximum Handgun Cartridge Pressure Specifications

Caliber	Pressure (psi)
.32 ACP	20,000
.357 Magnum	34,300
38 S&W	13,200
.38 Special	17,000
.38 Super +P	35,500
38 S&W	13,500
.38 Special	17,000
10-mm Auto	37,000
9-mm Luger	34,000
9-mm Luger +P	37,500
22 Long Rifle (rimfire)	23,500
.22 Short (rimfire)	22,000
.22 WRF	18,000

TABLE 11.4
Approximate Average Maximum Shotgun Cartridge Pressure Specifications

Caliber	Pressure (psi)
410 Bore 2 ½"	13,000
.410 Bore 3"	14,000
10 gauge	10,000
12 gauge (all but 3 ½" Magnum)	12,000
12 gauge 3 ½" Magnum	13,000
16 gauge	12,000
20 gauge	11,500
28 gauge	13,000

Note: For exact values refer to the SAAMI specification.

12 Interpreting the Technical Specification

OBJECTIVES OF INTERPRETING TECHNICAL SPECIFICATION

Interpretation of technical specification is necessary to select a firearm to meet the criteria in the given order of precedence:

- Safety
- Reliability
- Combat quality
- Minimum maintenance
- Minimize life cycle cost

INFANTRY WEAPONS OF SMALL ARMS CLASS

- Assault rifles
- Submachine guns
- Sniper rifles
- Light machine guns
- Shotguns
- Marksman rifle
- Handguns
- Heavy machine gun

The range of small arms ranging from caliber 4.73 onwards up to 20 mm are extremely wide and when combined with different architectures and operating mechanisms results in several categories and variants. Thus it presents an extremely complex scenario for choosing an appropriate weapon. Therefore, attempt has been made to mention only a few weapons as given below.

- **Assault rifles/ carbine:** M4A1 carbine, M16 , FN FAL , SIG MCX , AK-47, SA80 FN SCAR
- **SMGs:** MP7, MP5, Uzi, PP19 Bizon, Steyr Aug, FN-P90
- **Sniper:** Dragunov, AS-50, Barrett XM109
- **Shotguns:** Benelli M4, Franchi SPAS 12, Mossberg 500
- **MMG:** PKM, MG34, M240
- **Marksman rifle:** HK DMR, FN SCAR
- **Handgun:** Glock 19, M1911, S&W M28, Desert Eagle, Hi- Point Model JHP

DOI: 10.1201/9781003199397-12

Some weapons are claimed as the best. Let us examine.

1. **M4A1 (Assault Rifle)** (USA) – massively accurate with a fast fire rate.

FIGURE 12.1 M4A1.

   It is a Direct Impingement (DI) gas-operated weapon. Weighing only 7.5 lb. with 30 rounds of 5.56 × 45 mm NATO ammunition. It fires at a rate of 900 RPM, with a muzzle energy of 1,750 J with a 55 gr FMJ bullet. The flip side is that it fouls the receiver in sustained firing, and 55 gr bullet's energy is not very easily amenable for volume fire at the rated cyclic rate. See Figure 12.1.
2. **MP7 (Submachine Gun) – A mix of design of AK 47, M 16A1 action & Glock trigger mechanism –** Germany
   It is safe and beat out assault rifle users with relative ease in close quarter battle, which means the huge number of M4A1 users will not be able to match the firepower of the MP7. It is a Short-Stroke gas-operated weapon. Weighing 4.63 lb. with 20-round magazines. It can also accept 30, 40, and 50 detachable box magazines of 4.6 × 30 mm HK ammunition. It fires bullets of 26–42 gr at a rate of 950 RPM, with a muzzle energy in the range 447–540 J, making it extremely amenable for bullet spraying on mass target. The con is a complete polymer encloser that has a poor arrangement for heat dissipation in the volume fire. See Figure 12.2.

FIGURE 12.2 MP7.

3. **PKM (Medium Machine Gun)** – Soviet Union
   PKM is one of the most versatile guns in the game, claimed to be able to spar with snipers and assault rifles alike. It is an LS (Long Stroke) open bolt gas-operated weapon. Its weight is 16.53 lb. It is fed by 50 rounds of disintegrating metal belts of 7.62 × 54 mmR at a rate of 650 RPM. It fires bullets ranging from 150 gr to 181 gr with a muzzle energy of 3,629–3,779 J. It gives a cover firing up to a range of 1,094–1,640 yd. The extreme high muzzle energy does not allow it to be used on assault rifle role, though it is claimed so as mentioned above. See Figure 12.3.

FIGURE 12.3 PKM.

4. **MP5 (Submachine Gun)** – Germany
   Shooters will never struggle in close quarters (except against a shotgun). It is a roller-delayed closed bolt action weapon keeping the receiver extremely clean and free from operating gases. It weighs from 5.5 to 7.5 lb. in several versions. It fires 9 × 19 mm Parabellum (115–124 gr, muzzle energy 481–617 J), 10 mm Auto (135–230 gr, muzzle energy 880–1,041 J), and 0.40 S&W (115–200 gr, muzzle energy 680–790 J) in 15/30/32/40 round detachable box magazine or 100 rounds Beta C drum magazine. It fires at a rate of 700–900 RPM. It can be appreciated; the higher energy bullet though provides more lethality but makes the weapon increasingly difficult in volume firing in the assault mode. So, the selection of weapons is of utmost importance for the mission objective. See Figure 12.4.

FIGURE 12.4 MP5.

5. **Steyer AUG (Bullpup rifle)** – Austria
   It shoots very fast and deals solid damage at all ranges. It is a SS (Short Stroke) gas-operated closed bolt bullpup rifle. It weighs around 7.3–7.9 lb. It fires 0.223 REM 55 gr ammunition with a muzzle energy of 1,715 J. It has an effective range of 980 ft. Its high muzzle energy limits its volume of fire which is rated at 650 rounds per minute. Though it is a short-stroke action because of integrated plastic molding of the frame, its heat management is quite challenging, and its endurance in sustained firing is not known. The advantage of the weapon is this it is equipped with a single-stage progressive trigger and doesn't need a firing mode selector switch. See Figure 12.5.

FIGURE 12.5 Steyer AUG.

6. **AK-47** – Soviet Union
   The AK-47 is a high damage weapon but demands high skill and physical strength, requiring the shooter to control difficult recoil at a high rate of fire. It is an LS (long-stroke) closed-bolt gas action weapon. It weighs around 7.7 lb. It fires 7.62 × 39 mm ammunition. Its long stroke integral piston carrier mechanism causes a rapid change of a significant measure of the position of the center of mass of the weapon system. This induces an imbalance in the weapon hold and is responsible for its infamous inaccuracy and uncontrollability. The muzzle energy is ranging from 2,056 to 2,179 J with different bullet grains from 122 to 154 gr. This high muzzle energy generates almost uncontrollable recoil and keeping aim during firing is difficult. That's why there's a switch from 7.62 × 39 mm to 5.45 × 39 mm in the latest versions of AKs. However, the weapon has proliferated so much, and its ammunition is available in abundance, that it is present in almost every corner of the world, and people in possession want to retain it as a jealous prize. See Figure 12.6.

FIGURE 12.6 AK-47.

7. **FN SCAR H** – Belgium/America
   It's a SS (Short-stroke) closed bolt gas action weapon making the receiver free from gas fouling. It fires 7.62 × 51 mm NATO cartridge. The standard edition weighs around 7.9 lb. The muzzle energy of this rifle is 3,470–3,560 J firing bullets of 147 gr and 175 gr, respectively. The fantastic muzzle energy can take foes down over a long range of around 550–660 yards. It can be managed well in the hand of a capable wielder from a long distance but it is neither suitable nor manageable in close-quarter combat because of high recoil and uncontrollability. Its rate of fire of 625 RPM remains a theoretical capability. See Figure 12.7.

FIGURE 12.7 FN SCAR H.

8. **FAMAS F1** – France
   The FAMAS has three-round burst control but is ineffective beyond the mid-range, at which point the weapon is no mode effective. It is a delayed blowback bullpup assault rifle. It has a complex mechanism for action and hence poses a proportional problem in maintenance. It fires 5.56 × 45 mm NATO brass casing ammunition. The rifle weighs around 7.96 lb. It has a muzzle energy of 1,797 J when firing a bullet of 62 gr (SS109). It has an

effective range of 328.08 yards for the F1 version and 492.12 yards for the G2 version. Its endurance strength on sustained firing is not known. The total plastic body is a perfect heat cavity and the mechanism of its cooling on firing is very slow. See Figure 12.8.

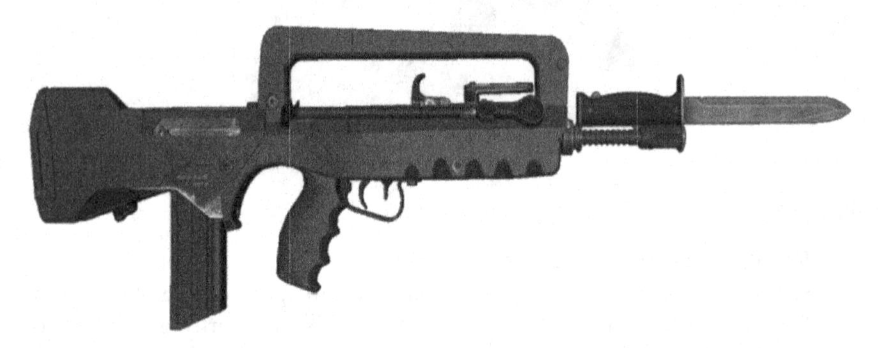

FIGURE 12.8 FAMAS F1.

9. **FN FAL** – Belgium
   It is one of the powerful weapons in terms of lethality and range. It is a short-stroke gas-operated assault rifle. It fires 7.62 × 51 mm NATO rounds. It weighs 9.8 lb. It has a muzzle energy of 3,470 J when firing a bullet of 147 gr. Its effective range is around 656.16 yards. The muzzle energy is such a high value that it produces a recoil of 130 J/s, making it impossible to control the fire at its rated cycle of around 700 rounds per minute. It's a rifle unsuitable for the CQB role. Its construction is heavy and the frame is milled out of solid forging, and the weapon is expensive. See Figure 12.9.

FIGURE 12.9 FN FAL.

10. **Dragunov** – Soviet Union
    The Dragunov trades off damage in favor of the fire rate. It can kill with a headshot. Though the semi-automatic nature of the gun allows for a follow-up, it simultaneously compromises with the desired accuracy of a sniper. It is a short-stroke gas-operated sniper rifle. It fires 7.62 × 54 mmR cartridge. Its unloaded weight without scope is 9.48 lb. Its muzzle energy is 3,614 J when firing a bullet of 181 gr. It has an effective range of 874.89 yards. As a sniper, it has a comfortable recoil of 13.8 J per shot. See Figure 12.10.

FIGURE 12.10 Dragunov.

11. **L86 (LSW) –** The United Kingdom
    It is one of the weapons in the L85 A2 series comprising of the assault rifle, LSW (Light support weapon), and Carbine. This is a magazine-fed squad automatic weapon intended to provide fire support at a fireteam level. The weapon has a heavier, longer barrel compared to the rifle and features a shorter handguard with an integrated bipod protruding from the front. Its rate of fire is 610–775 RPM. It is a short-stroke gas-operated LMG. It fires 5.56 × 45 NATO cartridges. Its weight with loaded magazine and sight attachment is 14.5 lb. It has a muzzle energy of 1,797 J when firing a bullet of 62 gr (SS109). It has a length of 35.4". Its effective range is 874.89–1,093.61 yards. The L86 was a target of criticism due to its inability to deliver sustained automatic fire as it lacks a quick-change barrel and belt feed. The apparent advantage is that it is a member of a family in a common series. See Figurc 12.11.

FIGURE 12.11 L86(LSW).

12. **The AK-74:** Kalashnikov – Soviet Union

FIGURE 12.12 AK-74.

It is the standard Russian infantry rifle. AK-74 came much later in the service of the Russian army. It replaced the AK-47 and AKM. It operates on the same principle as that of AK-47. It fires 5.45 × 39 mm ammunition replacing 7.62 × 39 mm cartridge. Thus, it settled for a muzzle energy of 1,400 J in place of more than 2,100 J of 7.62 × 39 mm cartridge. The reason is that burst firing is almost uncontrollable using 7.62 × 39 mm cartridge, and lower power cartridges (nearly 50% less power) address this issue very well. See Figure 12.12.

Note: Recoil energy analysis and evaluation of ease of firing in automatic firing mode of automatic weapons are very important while shooting a weapon. The recoil energy plays a vital role because if the recoil energy is not managed properly, then it will be difficult to fire the weapon for the shooter. The analysis of the recoil energy of some of the weapons is given in Table 12.1.

REPUTED SMGS FOR CLOSE QUARTER COMBAT

Uzi (SMG) – Low damage is made up of a fast rate of fire and high mobility, but outside of indoor combat, you'll find little use for this gun.

P90 (SMG) – Superfast fire rate, at the cost of almost everything else. This gun won't cause much damage at anything above point-blank range, and even then, there will be better options.

SOME HANDGUNS OF REPUTE

- Glock 19 – It is a 9 mm polymer frame short-recoil semi-automatic pistol effective for firing up to a range of 50 m.
- M1911 – It is a 0.45 caliber semi-automatic short recoil single-action pistol, having an effective range of 50 m. It is also referred to as colt 1911.
- S&W M28 – It is a double-action, 6 round cylinder revolver that fires a 0.357 magnum cartridge. It is manufactured with variants of barrel lengths from 4 inches to 8 inches.
- Desert Eagle – It is a semi-automatic pistol using a gas-operated ejection and chambering mechanism. It fires the most powerful 0.50 Action express cartridges. It has a firing range of up to 200 m.
- Hi-Point Model JHP – It is a 0.45 caliber, semi-automatic pistol. It uses a polymer frame. It is mentioned here for comparative study because of its blowback action as opposed to the short recoil operation that is usually used by other manufacturer's pistols and its often criticized performance.

TABLE 12.1
Recoil Energy Analysis and Evaluation of Ease of Firing in Automatic Firing Mode

SL No.	Weapon	Mass (lb) M	Bullet Type	Muzzle Energy (J) M.E.	Recoil Energy (J) R.E. $K_f \times \left(\frac{m}{M}\right) \times ME$	K_f	Rounds per minute (RPM)	Recoil Energy Transmit per second. R.E. × RPM
1	M4 Carbine	7.75	5.56 × 45 mm XM193 FMJ	1,755	3.38053761	1.9	900	50.70806419**
2	HK MP7	4.63	4.6 × 30 mm CPS	447	0.681274341	1.9	900	10.21911512
3	PKM MAG	16.53	7.62 × 54 mmR FMJ	3,614	8.653726279	1.53	650	93.74870136**
4	HK MP5	6.8	9 × 19 mmP Federal FMJ	481	1.742210294	1.5	950	27.58499632*
5	Steyr AUG	7.9	0.233 REM	1,265	2.390417658	1.9	700	27.88820601*
6	AK 47	9.4	7.62 × 39 FMJ	2,108	5.976987319	1.53	600	59.76987319**
7	Scar-L STD	7.3	7.62 × 51 mm FMJ	3,470	14.97328767	1.5	625	155.9717466***
8	Famas F1	7.96	5.56 × 45mm SS109 FMJ	1,797	3.799052148	1.9	1,000	63.3175358**
9	FN FAL	9.8	7.62 × 51 mm FMJ	3,470	11.15357143	1.5	700	130.125***
10	Dragunov	10.3	7.62 × 54 mmR FMJ	3,614	13.88797043	1.53	30	#
11	L86	14.5	5.56 × 45 mm SS109 FMJ	1,797	2.085548628	1.9	750	##
12	AK74	6.8	5.45 × 39mm	1,328	2.7649292	1.87	800	36.86572267*
13	QBZ95-1	7.2	5.8 × 42 mm DPB10	1,926	6.29468374	2.32	650	68.19240718**

The technical characteristics indicated in the table are indicative and approximate. The reader may refer to the manufacturer's data sheet for exact values.
***Impossible to fire in auto mode, **Very difficult to fire in auto mode, *Moderately difficult to fire in auto mode.
#Sniper semi auto, ##LSW- Light Support Weapon/LMG.

COMPARING SPECIFICATION OF ASSAULT RIFLES (TABLES 12.2 AND 12.3)

TABLE 12.2
Comparing Assault Rifle Specifications

Firearm	AK-47	M16A1
Weight (with loaded 30-round magazine)	4.78 kg	3.6 kg
Overall length	87.0 cm	99.0 cm
Barrel length	40.6 cm	50.8 cm
Cartridge	(M43) 7.62 × 39 mm	(M193) 5.56 × 45 mm
Muzzle velocity	(710 m/s)	(990 m/s)
Effective range	(350 m)	(460 m)
Accuracy @ 100 meters	(15 cm)	(11 cm)
Rate of fire	600 rounds/min	700–950 rounds/min
Standard magazine capacity	30 rounds	30 rounds
Rate of fire	600 rounds/min	700–950 rounds/min

The technical characteristics indicated in the table are indicative and approximate. The reader may refer to the manufacturer's data sheet for exact values.

TABLE 12.3
Differences in M16 & AK-47

M16	AK-47
1. Direct impingement gas-operated, magazine-fed rifle, with a rotating bolt and straight-line recoil design.	Works with long-stroke-piston operated by gas, with a rotating bolt, and fed by box/drum magazine, designed to be a simple, reliable automatic rifle.
2. Lightweight assault rifle, and to fire a new lightweight, high-velocity small-caliber cartridge.	A reliable automatic rifle that could be manufactured quickly and cheaply, using mass-production methods.
3. It was designed to be manufactured with the extensive use of aluminum and synthetic materials by the state-of-the-art computer numerical control (CNC) automated machinery.	Its receiver is designed to be stamped from sheet metal with a milled trunnion insert. It has many components/parts produced by investment casting.
4. Receivers may also be made from titanium and a variety of other metallic alloys, composites, or polymers.	They are also made with the use of synthetic/plastic furniture, such as folding stocks, handguards, and pistol-grips.
5. The AR-15/M16-series rifles are considered the finest human-engineered assault rifles in the world.	The M16 is ergonomically superior to the AK-47 in most respects. With the AK-47, the safety is a large lever not at all easy to manipulate.

(continued)

TABLE 12.3 (CONTINUED)
Differences in M16 & AK-47

M16	AK-47
6. M16 is a modular weapon system, easily configured as an assault rifle, a carbine, a submachine gun, and an open-bolt squad automatic weapon.	AKs come in 7.62 × 39 mm (AK-103), 5.45 × 39 mm (AK-74M), and 5.56 × 45 mm (AK-101), with cold hammer-forged barrels.
7. The (M16's) Stoner system provides a very symmetric design that allows the straight-line movement of the operating components.	With the AK-47's long-stroke piston gas system, the piston is mechanically fixed to the bolt group and moves through the entire operating cycle.
8. This allows recoil forces to drive straight to the rear. Instead of connecting or other mechanical parts driving the system, high-pressure gas performs this function, reducing the weight of moving parts and the rifle as a whole	The main disadvantage of this system is that the point of aim is disturbed due to change in the center of mass of the weapon during cyclic action which also results in changes in the force of recoil due to the sudden stop of the bolt carriers at the beginning and at the end of the cyclic travel.
9. Free recoil momentum	
40.4 lb.-fps	54.3 lb.-fps

The technical characteristics indicated in the table are indicative and approximate. The reader may refer to the manufacturer's data sheet for exact values.

COMPARING LMG SPECIFICATIONS (REFER TO TABLE 12.4)

- The LMG plays an infantry support role. It is designed to be deployed by an individual soldier, with or without an assistant.
- Uses the same ammunition as an assault rifle or may use full-power 7.62 × 51-mm cartridge when intended for use in the MMG role.

MMG

- The MMG is a fully automatic machine that uses full-power rifle-caliber ammunition.
- **Western** MMG/GPMG weapons, as on date, are designed to fire 7.62 × 51 mm full-power rimless rifle ammunition.
- **Eastern** MMG/GPMG weapons are designed to fire 7.62 × 54 mm R full-power rifle ammunition with a rimmed cartridge.
- FN MAG (as the M240 machine gun), which is generally called the M240 medium machine gun.

Comparing the specifications of MMG-M60 Western block weapon and PK machine gun (MMG) – Eastern Bloc weapon Refer to Tables 12.5 and 12.6.

TABLE 12.4
Differences between M249 & Bren MK4 Machine Guns

LMG Specifications	M249	Bren (Mk4)
Mass	• 7.5 kg (17 lb.) empty • 10 kg (22 lb.) loaded	• 9.75 kg (21.6 lb.) loaded
Length	40.75 in (1035 mm)	42.9 in (1,156 mm), Mk IV
Barrel length	• 465 mm (18 in) • 521 mm (21 in)	• 25 in (635 mm)
Cartridge	5.56 × 45 mm NATO	7.62 × 51 mm NATO (post-WWII)
Action	Gas-operated long-stroke piston, open bolt	Gas-operated, tilting bolt
Rate of fire	725-rounds per minute (r/min) or 1,000 r/min	500–520 rounds/min
Muzzle velocity	915 m/s (3000 ft/s)	2440 ft/s (743.7 m/s
Effective firing range	• 700 m • 3,600 m (maximum range)	• 600 yd (550 m) • 1,850 yd (1,690 m) max
Feed system	M27 linked disintegrating belt, STANAG magazine	20-round L1A1 SLR magazine 30-round detachable box magazine 100-round detachable pan magazine

The technical characteristics indicated in the table are indicative and approximate. The reader may refer to the manufacturer's data sheet for exact values.

TABLE 12.5
Specification of MMG-M60 Western Block Weapon

Cartridge	7.62 × 51 mm NATO
Caliber	7.62 mm
Action	Gas operated, short stroke, open bolt
Rate of fire	550–650 r.p.m.
Muzzle velocity	853 m/s
Effective firing range	1100 m
Feed system	Disintegrating belt with M13 Links
Sights	Iron sights
Mass	10.5 kg
Length	1105 mm (43.5 in)
Barrel length	560 mm (22.0 in)
Max. effective range	1,860 yd (1700 m)

The technical characteristics indicated in the table are indicative and approximate. The reader may refer to the manufacturer's data sheet for exact values.

Design Outlines of PKM

Operating mechanism – The bolt and carrier design is similar to that of the AK-47, oriented upside down compared to the AKM, the piston, and gas system underneath the barrel.

TABLE 12.6
PK Machine Gun (MMG) Eastern Block Weapon

Rate of fire	PK, PKM: 650 rounds/min Practical: 250 rounds/min
Muzzle velocity	825 m/s (2,707 ft/s)
Effective firing range	1,000–1,500 m sight adjustments
Maximum firing range	3800 m
Feed system	Non-disintegrating 50-round metallic belts in 100- and 200/250-round ammunition boxes
Sights	Tangent iron sights (default); optical, night-vision, thermal, and radar sights
Mass	7.5 kg (gun + integral bipod) + 4.5 kg (tripod)
Length	1,192 mm (46.9 in)
Barrel length	605 mm (23.8 in) (without muzzle device)
Cartridge	7.62 × 54 mmR
Action	Gas-operated, open bolt

The technical characteristics indicated in the table are indicative and approximate. The reader may refer to the manufacturer's data sheet for exact values.

Cartridge – The rimmed 7.62 × 54-mmR cartridges in a metal ammunition belt and held against the shoulder inside non-disintegrating looped links, leaving the rim exposed at the rear.

Receiver – U-shaped receiver is stamped from a smooth 1.5-mm (0.06 in) sheet of steel that is supported extensively by pins and rivets.

Barrel – A detachable barrel assembly that slides into the receiver and is fixed by a barrel-lock. The bore is chrome-plated and it has four right-hand grooves at a 1 in 240 mm (1 in 9.45 inches) rifling twist rate. The threaded muzzle can hold various muzzle devices such as flash hiders and muzzle brake.

Adjustable iron sights – 100 to 1,500 m. With a sight radius of 663 mm.

Optical sights – Warsaw Pact side-rail bracket on the left side of the receiver can mount various aiming optics.

Trigger – Operated by the mainspring and suitable for automatic fire. It has no single-shot mode.

Stock – The buttstock is skeletonized, provided with a pistol grip and a foldable carrying handle on the barrel.

Feed Mechanism

- The feed mechanism of the PK machine gun consists of two stages.
- It has a lever-type feed mechanism, having a feed pawl and roller.
- It first pulls the 7.62 × 54 mmR rimmed cartridge and then drops the cartridge down into the feed way. The bolt strips and feed the cartridges into the barrel chamber for firing.

GAS OPERATED (FIGURE 12.13)

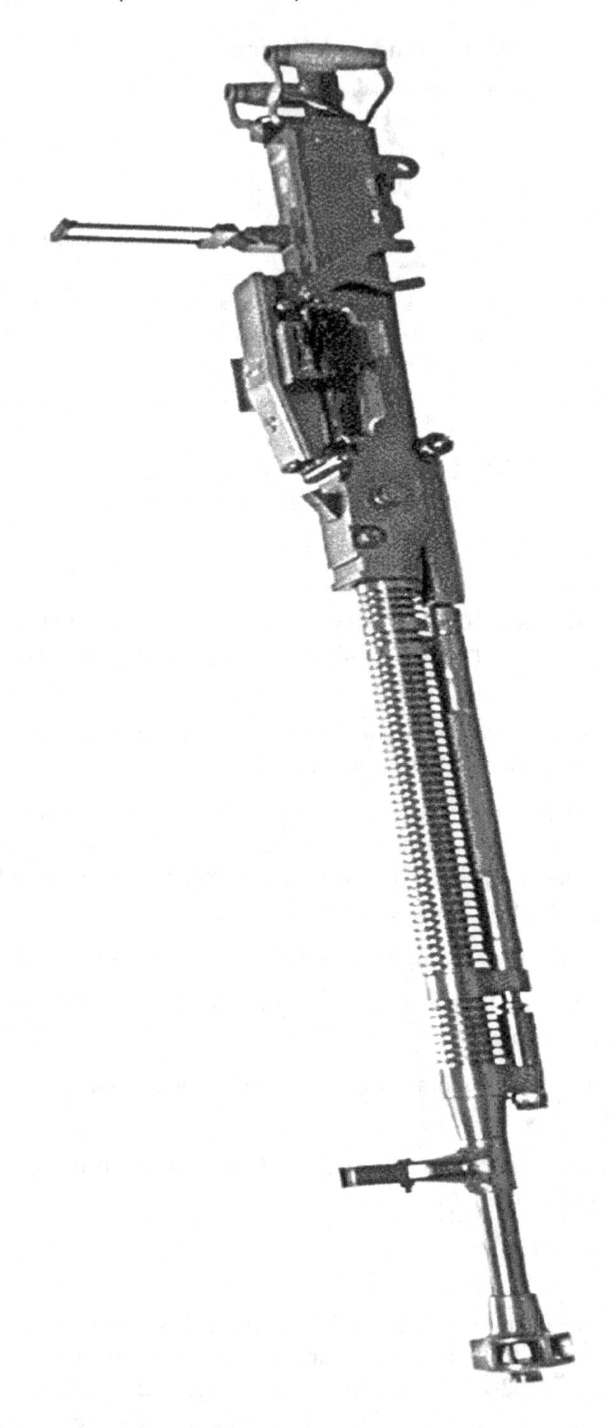

FIGURE 12.13 HMG – Heavy machine gun caliber 12.7 × 108 mm.

RECOIL OPERATED (FIGURE 12.14)

FIGURE 12.14 HMG – Heavy machine gun Browning M 2 .50 CAL.

Classical Specification (12.7 mm HMG)

When integral close-range support is needed, HMG from a vehicle/ground mount tripod is the best option. It is deployed against unarmored or lightly armored vehicles, infantry and boats, low-flying aircraft and, light fortifications. Refer to Table 12.7.

TABLE 12.7
Specifications of 12.7 mm Heavy Machine Gun

Effective range	Up to 2000 m
Caliber/cartridge	12.7 mm ×108 mm
Muzzle velocity	915 m/s
Length	1,656 mm
Barrel length	1,143 mm
Weight	38.15 kg (gun only)
Cyclic rate of fire	485–635 rpm
Feed system	Belt-fed with 50-round belts (detachable box magazine)
Weight of weapon	34 kg (without a tripod), 58 kg (with tripod)

The technical characteristics indicated in the table are indicative and approximate. The reader may refer to the manufacturer's data sheet for exact values.

What Else Is Provided in the Specs?

- Gas operated/recoil-operated
- Air/water-cooled machine gun
- Fires from an open bolt/closed bolt
- Selective/automatic mode only
- The gas piston and chamber location
- Type of the gas piston the long-stroke/short stroke
- A gas regulator is provided to adjust the rate of fire
- Recoil management arrangement (e.g. spring buffers, for the bolt and the bolt carrier)
- The cooling arrangement/barrel construction
- Muzzle brake
- The barrel replacement facility

What's the Normally Offered Package?

Specimen HMG Case

- Gun with breech assembly and muzzle brake
- Ammunition type compatibility
- Five pieces of standard ammo boxes with 50-round belts
- A piece of movable and adjustable weapon support
- One piece of adjustable carriage folding tripod/rotating column

- One piece of water-proof protective gun cover and barrel cap
- One piece of cleaning-kit and tool-kit carrying bag
- A small number of spare parts and tools maintenance kit
- One piece of standard cleaning kit including cleaning rod, brush, oilcan, and rag

What Are the Missing Items in the Specification Package?

- Construction details and materials used to manufacture frame, barrel, and actions
- Trigger mechanism types and safety features
- Barrel change round count
- Severely stressed changeable parts round counts (e.g., buffer/recoil spring, extractors/extractor spring, firing pin/striker, barrel, hammer/striker
- Details of firing mechanism including bolt and bolt carrier
- Information on design complexity of feed mechanism in M13 link belt feed system
- Firearms adaptability according to weather and terrain conditions
- Surface and heat treatment conditions of various parts
- Firearms approximate life
- Meltdown counts
- Barrel life
- Detail design features of parts with the shortest life
- Information on special surface/metallurgical treatment for enhancement of essential functions
- Cookoff-limit in case of closed bolt actions
- Heat management, particularly in Bullpup design – the biggest challenge for sustained firing

Design Details and Features to be Examined Answered

- Does it fire from a closed bolt/open bolt position?
- Is it operated on the short recoil/gas operation principle?
- Does it have varying cyclic rates of fire to fully automatic? Can it be selected to fire single-shots?
- What are ammunition types compatibility?
- Does it include M33 type ball (706.7 grain) for personnel and light material targets? M17 type tracer?
- Can it fire M8 type API (622.5 grain), M20 API-T (619 grain), and M962 SLAP-T, M903 SLAP (Saboted Light Armor Penetrator) to perforate 1.34–0.75 inches (34–19 mm) of FHA (face-hardened steel plate) at ranges of 500–1,500 m and beyond?
- Is any blank-firing adapter (BFA) of a special type must be used as a recoil booster to allow the recoil-operated action to cycle when firing blanks during peacetime operations?

- What are the variants and derivatives (12.7 × 99 mm) available and where is it placed in comparison?
- Is it air/water-cooled?
- How much recoil in the case of rifles and handguns?

AUTHORS' NOTES ON PRESENT-DAY FIREARMS

Most Significant Improvement

- Integration of bullet, propelling charge, and means of ignition in a single unit- cartridge
- Qualitative improvement directed towards propellant–ignition mechanism
- Gun metallurgy/material
- The transition from metal to plastic
- Rifling – sighting system, reliability, and efficiency – effectiveness and accuracy

Plastic Injection Molding (PIM) can Improve the Form, Function, and Fitness of the Weapon

- Form: Proven processes can improve aesthetics by eliminating issues like visible seams and stress, uneven and unattractive texture, and colors.
- Function: The best-suited firearm materials should be resistant to heat scratch and impact. The right mix of material in PIM firearms can be constructed with the desired strength and favorable physical and chemical properties. Nylon Type 6 and Type 6/6 are the candidate materials.
- Fit: Interchangeability and smoothness in the assembly are the most desirable characteristics of the firearm system. PIM gives a definite lead in this matter in manufacturing firearms and results are better assemblies and better productivity.

Polymers are durable, resistant to corrosion, and lightweight, and they can be molded to any shape and are not so expensive. So, weapons with polymer frames are much lighter and concealable to carry. The polymers have more flexibility and less rigid than steel and hence more compressible. Therefore, when the gun is fired, a polymer frame can absorb significant recoil energy and minimize the perceived recoil. Thus, polymer has become a popular material in the construction of firearms.

Sighting System

- There are major improvements with optical sights, laser beam sights, and other digital devices for correctly estimating the point of aim and target range.
- Provisions for correction of wind speed and correction for Earth's rotation.

Two Present-day Good Quality Firearms

Assault Rifle

1 HK G36E (Heckler & Koch) – Germany

FIGURE 12.15 HK G36E.

The Heckler and Koch G-36 assault rifles were produced as an HK-50 project in the 1990s. Its importance can be understood from the fact that its kinetics was once considered as a basic platform for the canceled US XM-29 OICW weapon and an XM8 assault rifle project. The modularity of the weapon is of superb class and can be assembled and disassembled without major tools. It is ambidextrous and fires in single-shot 2-round burst and full-auto modes. It fires 5.56 × 45 mm ammunition (Figure 12.15).

Precession Rifle

FIGURE 12.16 Ruger Precision Rifle (RPR) – U.S.A.

Ruger precision rifle is a modular, bolt-action rifle promising sub-MOA (less than one MOA) performance in the U.S was manufactured by Sturm, Ruger & Co. in 2015. The Precision Rifle was designed to maintain an in-line recoil path. Perceived recoil is directed back straight through the buttstock. It has an effective range of 1,600 yards and a maximum range of up to 2,000 yards. It is designed in 5.56 and .308 caliber barrel (Figure 12.16).

A Unique Rifle

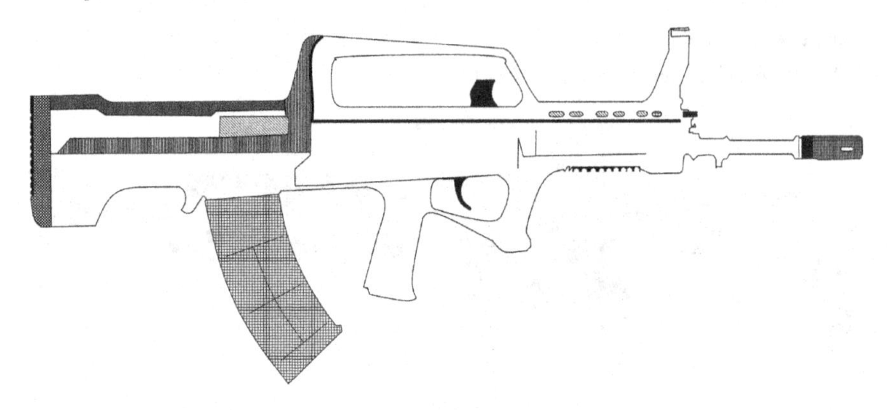

FIGURE 12.17 QBZ 95-1 – China.

QBZ 95-1 is a member of the type 95-gun family with a Bullpup configuration similar to the British SA 80. It is an assault rifle. The acronym QBZ is explained as Q – Qing Wugi (light weapon), B – Buqing (rifle), Z – Zidong (automatic). It is a Chinese sobriquet of the queen of the battle that fires 5.8 × 42 mm DBP 10 ammunition at 650 rounds per minute, using a 30-round detachable box magazine. It weighs 7.2 lb. when empty. Its other cousins are QBB 95, LSW (light support weapon), and QBZ 95-B Carbine. Its export version is QBZ 97 that is designed to fire 5.56 × 45 mm ammunition. The weapon has many similarities with the French FAMAS, Austrian Steyer AUG, South African Vector CR 21, Singapore SAR 21, and Israel TAVOR.

The QBZ 95-1 is designed with a longer barrel (almost 20 in long), a muzzle brake, and a fore-handguard with a diamond-shaped cross-section to dissipate heat. The rifle is also equipped with a recoil buffer to manage the recoil. It uses a linear striker firing mechanism where a spring-loaded firing pin and a linear hammer fire the chambered cartridge. It is noteworthy as most military rifles use a rotating hammer firing mechanism. QBZ 95 uses inline main and striker springs that use the same guide rod instead of two parallel springs. It functions on the principle of the short-stroke gas piston. It has a four-position firing mode selector switch: 0 – safe, 1– semiauto, 2 – full auto, 3 – three-round burst setup. It can be equipped with a 35-mm UBGL (Under barrel grenade launcher).

It is a fierce competitor of the M16-A4 assault rifle (5.56 × 45 mm NATO), AK-74 assault rifle (5.45 × 39 mm Russia), and the Chinese claim that QBZ 95-1 outperforms not only these two but all other assault rifles of its class in terms of flatness of trajectory, penetrating power, and high downrange energy. Despite the particular claim, physics dictates that the rifle is troubled with excessive recoil and the efficacy of its shape and geometry from the ergonomics and heat transfer point of view are not beyond the scope of debate. The 5.8 × 42 mm ammunition is a steel case cartridge painted in dark brown color. For extraction reliability, the case has a thick rim and large extractor groove. The DBP 10 is a new cartridge with 71-grain FMJ bullets with a muzzle velocity of 915 m/sec and with a muzzle energy of 1,976 J. The DBP 10 has a copper alloy jacketed bullet with 3.8 mm diameter hardened steel core with G7BC of 0.193. The theoretical estimate of recoil is about 70 J/s. The degree of ease in automatic firing can be guessed from this (Figure 12.17).

New Challenges and Emerging Requirement

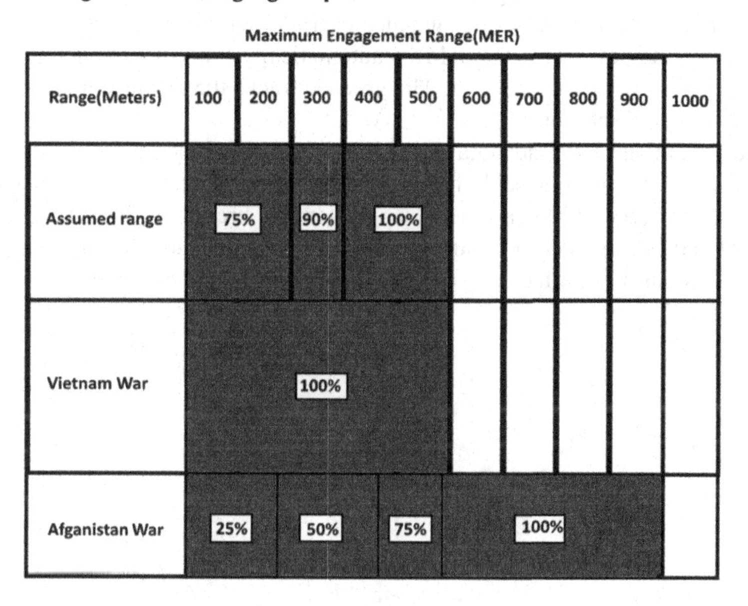

FIGURE 12.18 Maximum engagement range (MER).

The capability gap as in modern times can be appreciated from the maximum engagement range (MER) as observed by American troops in the Afghanistan War.

Figure 12.18 shows that the battlefield has not remained static and ever-changing with the type of enemies. No longer the scenarios of the Vietnam and Korean wars are valid where engagement about 90–100 % was taking place within a range of 500 m either by direct fire engagement or by ambush. The psychology of the emerging adversaries is fast-changing; they are making attacks from distant defilades and demand has cropped up for an engagement range of 800 m to defeat the enemy in their hideout. Thus, the need has arisen again to develop weapons based on smart ammunition.

The outcome of the Afghanistan war gave rise to new challenges. The weaponry supplied to the Afghan military came to the hands of the militants over the decades through battle losses or possibly by corruption. M4s, laser, optical scope, and the night-vision equipment got to the militants. They were no more required to fight in direct-fire or ambush strategy. They were enabled to fight from a distant defilade even in low visibility. So, counter defilade target engagement (CDTE) equipment became necessary to defeat the enemy. The equipment had to have a replacement of sophistication as a rifle with upgraded night vision and the optics such as 4X ACOG and the laser range finder and air bursting ammunition.

The development of XM25 is an endeavor to counter the emergent situation and is under the experiment to judge the efficacy.

XM-25 is designed to be a CDTE system. It is designed to be an individual semi-automatic system to fire programmed air-burst ammunition. The ammunition is a 25 × 40 mm grenade that can be launched over a programmable distance of 600 M with an accuracy of 3.0 m. The XM-25 automatically transmits the detonating distance to a grenade in the chamber. The grenade tracks the distance it travels by the count of its spiral rotation.

The weapon has its disadvantage. It's heavy and has a much-limited firing capacity. It has a 5-round detachable box magazine. It is not enough in a CQB role. A soldier has to carry an M4 carbine to supplement the firepower. But XM 25 together with M4, has been perceived as an objectionable weapon load (almost 25 kg) for the individual soldier. So, it has not yet filled the capability gap for the dream OICW.

Multiple Choice Questions for Practice

Q1. The first development of the firearm was in the form of:

a) Pistol
b) Cannon
c) Rifle
d) Machine gun

Q2. The basic of production of a firearm depends on harnessing:

a) Mechanical energy
b) Chemical energy
c) Sound energy
d) Electrical energy

Q3. The musket is a firearm with:

a) A smoothbore muzzleloader
b) A rifled muzzleloader
c) A rifled breech loader
d) Both rifled and smoothbore muzzleloader

Q4. Gun powder is a mixture of charcoal, sulfur, and —

a) Nitrocellulose
b) Potassium chloride
c) Potassium chlorate
d) Sodium nitrate

Q5. The firing mechanism evolved in the following sequence:

a) Flint lock, Matchlock, percussion lock, Wheel lock
b) Match lock, Flint lock, Wheel lock, Percussion lock
c) Flint lock, Percussion lock, Wheellock, Match lock
d) Match lock, Flint lock, Percussion lock, Wheel lock

Q6. The correct types of gun powder used by the early gunners are:

a) large-grained – cannon, medium-grained – shoulder arms, fine-grained – pistols
b) large-grained – pistols, fine-grained – shoulder arms, medium-grained – cannon
c) medium-grained – pistol, fine-grained – cannon, medium-grained – shoulder arms
d) large-grained – cannon, large-grained – cannon, medium-grained – pistol

Q7. The shape of the earliest gun barrel was similar to

a) A perfect cylinder matching constant velocity vs time curve
b) A sphere matching the impulse of explosion vs time
c) Matching the shape of a projectile velocity vs time curve
d) A shape matching the pressure vs time curve of an explosion

Q8. (Fill in the blanks) Three basic ideas based on which the firearm was developed as follows:

a) The first was that gunpowder's propellant force could be used most effectively by confining it within a ______________________ barrel.
b) The second perception was that methods of construction derived from cooperage could be used to construct ______________________ barrels. The third perception was that a ____________________ was the optimal projectile.

Q9. Standardized patterns and parts were developed to address the problems of

a) Inter-changeability
b) Mass availability
c) Better functionality
d) All of the above

Q10. In percussion ignition, the kind of thermo-chemical process that occurs in the modern cartridge is as under:

a) Burning at a constant rate and explosion
b) Deflagration with progressive burning
c) Detonation and Deflagration
d) Progressive burning and detonation

Q11. The modern form of development came up with the following sequences

a) Priming pellet, needle-shaped firing pin, steel cylinder called the bolt, the frame of the receiver.
b) The needle-shaped firing pin, the frame of the receiver, steel cylinder called the bolt, priming pellet.
c) The frame of the receiver, steel cylinder called the bolt, needle-shaped firing pin, priming pellet.
d) Priming pellet, needle-shaped firing pin, the frame of the receiver, steel cylinder called the bolt.

Q12. Fill in the blanks:

a) All early breechloaders used __________________ as their source of propellant energy.
b) The burnout efficiency of black powder is ______________ percent. Nitrocellulose produces ______________ times energy as compared to that of black powder.

Q13. Fill in the Blanks

a) ______________ was the first country to issue a small-bore high-velocity repeating rifle. Assault rifles operate by using either ___________ gases or __________ forces generated by a fired round to force back the bolt, __________ the spent cartridge case, and __________ the firing mechanism.

Q14. True or False:
a) Both MMG and LMG use an intermediate power cartridge.
b) MMG an HMG use full-powered cartridge.
c) LMG and Assault Rifle use an intermediate power cartridge.
d) SMG fires in a locked bolt condition.

Q15. The power of a firearm is most rapidly increased by increasing:
a) The depth of rifling
b) The length of the barrel
c) The bore diameter
d) The shape of cartridge case

Q16. The rifles are configured as Bullpup rifles by
a) Decrease in the length of the barrel and providing a scope
b) Placing the magazine ahead of the trigger
c) Keeping the barrel long and keeping the magazine behind the trigger
d) By providing a grenade launcher

Q17. The general principles of operation of firearms (match):

1. Short recoil operated	i. AK 74
2. Blowback	ii. Browning HP 9mm
3. LS gas operated	iii. M16
4. DI Gas operated	iv. MP5

Q18. Fill in the blanks:
a) The recoil operated firearms operates with ______________ bolt position.
b) The MMG and HMG operate______________bolt system to prevent______________.

Q19. The straight blowback firearms have a massive block breech to ensure:
a). Increase the strength of the block breech
b). To delay the rear motion
c). To match the mechanical impedance
d). Both B and C above

Q20. To have the optimum reliable design against fouling by the propelling gases the best design will be by the operation with:
a) DI
b) Blowback
c) LS gas (Expansion)
d) LS gas (Impulse)
e) SS gas (Expansion)
f) SS gas (Impulse)

Q21. What is the thermodynamic efficiency of a rifle when considered as an IC engine? (One line answer)

Q22. Between 0.303 Lee Enfield and 0.308 Win rifle which one is the most powerful and why? (One line answer)
Answer:

Q23. What is hang fire in an Assault rifle?
Answer:

Q24. What do you mean by stovepipe in small arms firing?
Answer:

Q25. What causes the double-feed problem in firing an assault rifle?
Answer:

Q26. The design for barrel strength gives the safest result by using the theory of:
a) Rankine
b) Haigh
c) St. Venant
d) Von Mises

Q27. The major elements of barrel geometry are chamber, bore, rifling, and ________________.

Q28. The length of the barrel should be sufficient:
a) To develop maximum peak pressure.
b) To achieve maximum muzzle velocity.
c) To ensure complete burnout.
d) All of the above.

Q29. The most thermo-mechanically stressed element of the gun barrel is:
a) Neck of the chamber
b) Throat of the chamber
c) Leade of the chamber
d) The rifling of the barrel

Q30. The maximum frictional force is developed in the barrel having
a) Conventional cut rifling
b) Cold forge land and groove rifling
c) Polygonal rifling
d) EDM cut rifling

Q31. An M4 carbine with a twist rate of 1 in 7 inches (177.8 mm) and a muzzle velocity of 3,050 feet per second (930 m/s) will give the bullet a spin of 930 m/s / 0.1778 m.
a) 256,000 rpm
b) 314,000 rpm
c) 350,000 rpm
d) 322,000 rpm

Q32. With a velocity of 600 m/s (2,000 ft/s), a diameter of 0.5 inches (13 mm), and a length of 1.5 inches (38 mm), The twist rate of rifling will be:
a) 23
b) 27
c) 25
d) 21

Q33. The strength of the sound of a gunshot or muzzle report is
a) 100 dB
b) 115 dB
c) 148 dB
d) 140 dB

Q34. The type of block breech used by FN FAL is
a) Rotating Block breech
b) Sliding Block breech
c) Tilting Block breech
d) Falling Block breech

Q35. A *striker* is essentially firing pin directly loaded to a ____________ eliminating the need to be struck intermediately by a ____________.

Q36. The ____________ interface between the trigger and the hammer/striker is typically referred to as the ____________.

Q37. A double-action, is a design which either has no ____________ mechanism capable of holding the hammer or striker in the ____________ position.

Q38. The AR-15/LR300 assault rifle has:
a) Release trigger
b) Preset trigger
c) Binary trigger
d) Variable trigger

Q39. The trigger pull can be divided into three mechanical stages: a)____________.b).____________.c)____________.

Q40. Metal injection molding (MIM) is a metalworking process in which ____________ is mixed with ____________ material to create a "feedstock" that is then shaped and solidified using ____________. The molding process allows high volume, complex parts to be shaped in a single step.

Q41. The MIM proper sequence is
a) Injection molding, Thermal debinding, Solvent debinding, Sintering, Mixing of substrate and binder
b) Mixing of substrate and binder, Injection molding, Solvent debinding, Thermal debinding, Sintering

c) Mixing of substrate and binder, Solvent debinding, Thermal debinding, Injection Moulding, Sintering
d) Mixing of substrate and binder, Solvent degreasing, Sintering, Injection Moulding, Thermal debinding

Q42. The optimum achievable strength of MIM parts is:
a) 40% of the original metal strength
b) 90% of the original metal strength
c) 98% of the original metal strength
d) 100% of the original metal strength

Q43. Which of the following is a modern debinder in the MIM process?
a) n-propyl bromide
b) trichloroethylene
c) 1,1,1- trichloroethane
d) perchloroethylene

Q44. The correct order of precedence in the importance of ammunition design
a) Safety – Reliability – weapon size and configuration – combat quality
b) Safety – weapon size and configuration – combat quality – Reliability
c) weapon size and configuration – combat quality – Safety – Reliability
d) Safety – Reliability – combat quality – weapon size and configuration

Q45. Internal ballistics consist of 1) ______________, 2) ____________, 3) __________ and______________ in a fire arms,

Q46. The acceptable failure probability in the small arms ammunition is of the order of:
a) 10^{-6}
b) 10^{-7}
c) 10^{-5}
d) 10^{-4}

Q47. Fill in the blanks:
The modern types of ammunition primer for metallic cartridges are ______________ and ______________.

Q48. Dragunov SVD sniper rifle uses:
a) Centerfire semi-rimmed cartridge
b) Centerfire rimless cartridge
c) Centerfire rimmed cartridge
d) Rimfire belted cartridge

Q49. The most reliable cartridge of assault rifle is:
a) Rimfire belted cartridge
b) Center fired rimless cartridge
c) Center fired semi-rimmed cartridges
d) Center fired rebated cartridge

Q50. From the increasing intensity of detonation, the precedence of the primer is:
a) Rimfire – Bedan – Boxer
b) Rimfire – Boxer – Bedan
c) Bedan – Boxer – Rimfire
d) Boxer – Bedan – Rimfire

Q51. The most important role of the case is feeding and ______________ of the breech.

Q52. Which of the following is the modern primer compound?
a) Fulminate of mercury
b) Chlorate compounds
c) Lead styphnate
d) Barium nitrate
e) Both a and c
f) Both c and d

Q53. For any given calibre and muzzle energy, there is an ________________ relationship between barrel length and the size of the ______________.

Q54. The inertia force for a bullet of 40 grains (2.6 g) to achieve a muzzle velocity of 900 m/s in transit time of 1.06 ms is
a) 3,412 N
b) 2,281 N
c) 2,418 N
d) 2,134 N

Q55. The observed velocity of a bullet of 7.62 × 51 mm ammunition fired through a 24-inch barrel is found to be 2,800 ft/s, average pressure in the barrel will be:
a) 60,000 psi
b) 18,700 psi
c) 55,000 psi
d) 8,000 psi

Q56. The minimum wall thickness of 12 bore smooth barrel gun can be as low as:
a) 0.25 mm
b) 0.10 mm
c) 0.55 mm
d) 1 mm

Q57. Which of the following weapon has a paraxial magazine?
a) Tavor X95
b) P90
c) AR 15
d) MP5

Q58. Given a supercavitation bullet, which one of the following weapons will be suitable for underwater warfare?
a) M16
b) SA80
c) AK 47
d) TAR 21

Q59. Which of the following muzzle devices manages both recoil, flash, and sound pressure?
a) Compensator
b) Flash hider
c) Muzzle brake
d) Suppressor

Q60. What is the maximum impurity level percentage permitted for sulfur and phosphorus in gun barrel material?
a) 0.01
b) 0.25
c) 0.025
d) 0.15

Index

Printed in the United States
by Baker & Taylor Publisher Services